Nanopores for Bioanalytical Applications
Proceedings of the International Conference

Nanopores for Bioanalytical Applications
Proceedings of the International Conference

Edited by

Joshua Edel
*Institute of Biomedical Engineering, Imperial College London,
London, UK
Email: joshua.edel@imperial.ac.uk*

Tim Albrecht
*Department of Chemistry & London Centre for Nanotechnology,
Imperial College London, London, UK
Email: t.albrecht@imperial.ac.uk*

RSCPublishing

The proceedings of the First International Conference on Nanopores for Bioanalytical Applications held in Lanzarote, Spain on 6-10th February, 2012.

Special Publication No. 340

ISBN: 978-1-84973-416-5

A catalogue record for this book is available from the British Library

Published by The Royal Society of Chemistry,
Thomas Graham House, Science Park, Milton Road,
Cambridge CB4 0WF, UK

Registered Charity Number 207890

For further information see our web site at www.rsc.org

Printed in the United Kingdom by Henry Ling Limited, Dorchester, DT1 1HD, UK

PREFACE TO THE PROCEEDINGS OF THE INTERNATIONAL CONFERENCE ON NANOPORES FOR BIOANALYTICAL APPLICATIONS

Conference Chairs: Tim Albrecht & Joshua Edel
                       Department of Chemistry
                       Imperial College London
                       SW7 2AZ, London, UK

The intention of organizing a conference dedicated to nanopores for bioanalytical applications was a result of wanting to showcase and highlight achievements made in this exciting topic over the past couple of years. Having a conference dedicated to this topic resulted in a unique opportunity to share information and scientific ideas amongst likeminded researchers. It is perhaps fair to say that this meeting brought together a group of world leading experts, both experimentalists and theoreticians, in the area of nanopore research. With this in mind, putting together proceedings dedicated to this conference was a natural thing to do.

While it was a large task editing the proceedings to this conference, we are delighted with the outcome of this publication. Working on such a project has truly been a privilege to us, and would therefore like to thank all the contributors. It is perhaps fair to say that without your contributions, this publication would not have happened. Although all the topics covered are based around nanoporous membranes, the depth of research covered is tremendous. We hope the papers in these proceedings will prove to be helpful towards highlighting all the interesting research taking place within the nanopore community.

Finally, we would like to thank the people behind the scenes at Zing Conferences. Without their support and backing, both the proceedings and the conference would have never taken place. Furthermore, we would like to thank the staff at the RSC for helping us with the administrative aspects and publishing of the proceedings.

April 29th, 2012

CONFERENCE ATTENDEES

About Zing Conferences

Zing Conferences was established by Mark Long in 2007 to serve as an opportunity for the global community of scientific activity and interest to meet, exchange ideas and stimulate fruitful collaborations. Set in glorious coastal locations, in luxury five-star accommodation, each conference aims to bring together a range of expertise, from Academia to Industry, from the Professional to the Student, and provide a beautiful and serene environment in which to present and discuss their respective work. Zing Conferences is the promotion of inter-disciplinary communication; of the successful cross-pollination of theories and research amongst the many allied scientific fields. To promote this free exchange and dissemination of new ideas and research, all delegates are invited to compete for a place in the main lecture programme, of some 40 speakers, or to present their work as part of the larger Posters sessions by submitting abstracts for Chair-lead consideration.

Each day is divided into a series of lectures, typically grouped around a unifying theme, and one of which also contains Poster sessions. The afternoon of each full day, meanwhile, is left free for each to enjoy any of the various activities offered by the Hotel, whether taking advantage of some of the water-sports on offer or simply spending a tranquil couple of hours on the beach or by the pool.

The content of each conference is entirely determined by the appointed Chairs, in most cases this will be a collaboration of 2 or 3 world renowned scholars. It is they who are responsible for the intellectual content of each programme, selecting and organising each schedule. Each speaker is allotted a generous time in which to present their work; Poster presentations are either given as a 4minutes Flash talk or an open, interactive Poster session.

Zing Conferences places no restrictions on attendance, encouraging participation from all walks of life and all levels of expertise.

Jane Hill

Director

Contents

IONIC CURRENT DETECTION OF DNA ORIGAMI NANOSTRUCTURES WITH NANOCAPILLARIES

N. A. W. Bell[1], S. M. Hernández-Ainsa[1], C. R. Engst[2], T. Liedl[2], and U. F. Keyser[1]

[1] Cavendish Laboratory, University of Cambridge, JJ Thomson Avenue, Cambridge, CB3 0HE, United Kingdom
[2] Center for NanoScience and Department of Physics, Ludwig-Maximilians-Universität München, Geschwister-Scholl-Platz 1, 80539 München, Germany

1 INTRODUCTION

A continuing challenge in the nanopore field is the control of geometry and surface functionality on the nanometre scale. Protein nanopores such as α-haemolysin can be genetically engineered to attach binding motifs at specific positions[1] but the narrow channel size limits most membrane proteins to the study of single stranded DNA and unfolded proteins. The mechanical instabilities of lipid bilayers and controlling the number of protein insertions also create difficulties for device integration. Synthetic nanopores made by silicon nanotechnology can be tuned to different diameters for the analysis of a wide range of biomolecules but the creation of fixed structures with atomic scale control has yet to be demonstrated. Hybrid nanopores offer a novel method for the formation of nanopores by directing a molecular construct into a silicon based nanopore[2]. This combines the advantages of a solid membrane support with a fixed number of pores and the ability to make atomically defined molecular constructs made by self-assembly.

We recently demonstrated first experiments on forming a hybrid DNA origami nanopore[3]. DNA origami is a well established technique for the formation of nanostructures with almost arbitrary geometry[4,5,6]. A 7-8kbp single strand of DNA is folded into a pre-programmed structure by adding 100s of short synthesised DNA strands known as 'staple' strands. Functional motifs can be added by standard biochemical modifications to the staple strands[7]. The process of design and synthesis of a DNA origami structure can be carried out in less than a week and several billion structures made simultaneously in high yields. This makes DNA origami ideally suited for the rapid prototyping of designer nanopore structures with the potential for massive parallelisation[8,9].

Here we describe the fabrication of a flat DNA origami nanopore structure. Our experiments show that translocations and trapping of these structures can be readily detected with nanocapillaries.

2 METHOD AND RESULTS

2.1 Fabrication of nanocapillaries

The fabrication of nanocapillaries and their use for detecting the folding state of DNA has been described in detail elsewhere[10]. Briefly, a laser assisted pipette puller (Sutter P-2000) is used to draw down the diameter of quartz capillaries to a few tens of nanometres. The nanocapillary is then glued into a PDMS mold so that it is the only connection between two fluid reservoirs. The device is plasma cleaned for five minutes before immediately adding buffered salt solution and desiccating to remove bubbles. These nanocapillaries are a simple, cheap alternative to nanopores made by ion beam milling methods with each nanocapillary costing less than $1 and taking less than a minute to produce.

Our laser pulling program produces a range of nanocapillary sizes of 29±8 nm as estimated from measuring 35 nanocapillary diameters by scanning electron microscopy[11].

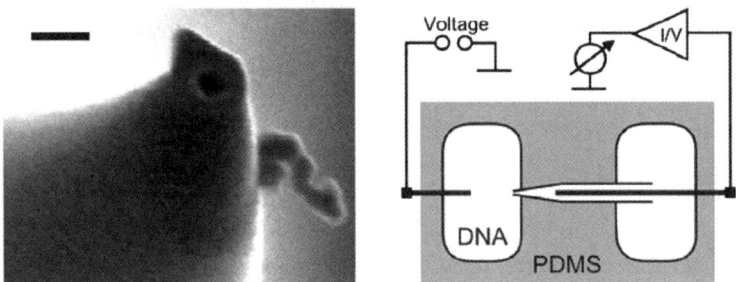

Figure 1 Left - SEM image of nanocapillary with approximately 3 nm coating of Pd/Au for imaging. Scale bar = 100 nm. Right - Schematic of measurement setup.

2.2 Synthesis of DNA origami

We designed a flat DNA origami shape approximately 60x50 nm² wide and two helices in height and with a 10x10 nm² nanopore in the centre. The structure was synthesised by mixing scaffold and staple strands in a 1:10 stoichiometric ratio in 1xTE buffer and 14mM $MgCl_2$ before heating to 65°C and cooling over 2 days. AFM imaging showed a high yield of the correctly folded structure.

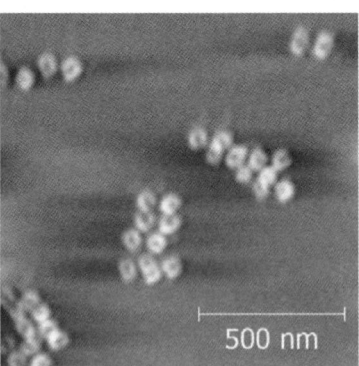

Figure 2 - Tapping mode AFM topographic image in air of DNA origami nanopore structures immobilised on mica.

2.3 DNA origami translocation events

For nanocapillaries with a resistance of less than approximately 20 MΩ we observe clear translocations when DNA origami is added to the negatively biased reservoir in a concentration of 0.5 ng/ L. Figure 3 shows typical translocation events. A bias of 500 mV was applied and the current signals were filtered at 30 kHz and sampled at 200 kHz. A solution of 0.5xTBE, 5.5mM $MgCl_2$, 1M KCl was used for all recordings.

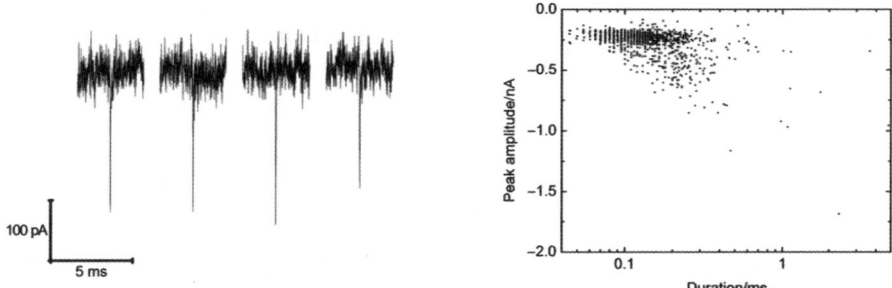

Figure 3 - Left - Example translocation signatures of DNA origami structures through nanocapillaries. Right - Scatter plot showing the duration and peak amplitude from 1003 translocation events.

2.4 Trapping of DNA origami
In capillaries with resistance greater than approximately 20MΩ, individual DNA origami structures can be trapped onto the end of the capillary. Figure 4 shows one such trapping event forming a hybrid DNA origami-nanocapillary nanopore.

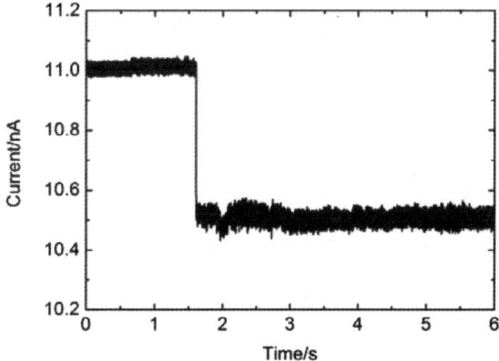

Figure 4 - Trapping of DNA origami nanopore on the tip of a nanocapillary. A bias of 200 mV was applied throughout the experiment and at 1.5 s a trapping event occurs.

3 CONCLUSION

DNA origami is a powerful technique for the construction of designer nanopores. Nanocapillaries can be used both to detect the translocation of DNA origami structures and to trap the origami to form a hybrid pore by tuning the diameter of the nanopore. The detection of translocations is a potential new tool for assessing the folding quality and aggregation of DNA origami in solution. The formation of a hybrid origami-nanocapillary nanopore represents a novel method for creating hybrid nanopores with high throughput and low cost.

We thank Lorenz Steinbock for SEM imaging of nanocapillaries.

References

1. S. Howorka, S. Cheley, and H. Bayley, *Nature biotechnology*, 2001, **19**, 636-9.
2. A. R. Hall, A. Scott, D. Rotem, K. K. Mehta, H. Bayley, and C. Dekker, *Nature Nanotechnology*, 2010, **5**, 874-877.
3. N. A. W. Bell, C. R. Engst, M. Ablay, G. Divitini, C. Ducati, T. Liedl, and U. F. Keyser, *Nano letters*, 2012, **12**, 512-7.
4. P. W. K. Rothemund, *Nature*, 2006, **440**, 297-302.
5. S. M. Douglas, H. Dietz, T. Liedl, B. Högberg, F. Graf, and W. M. Shih, *Nature*, 2009, **459**, 414-8.
6. T. Liedl, B. Högberg, J. Tytell, D. E. Ingber, and W. M. Shih, *Nature Nanotechnology*, 2010, **5**, 520-524.
7. N. V. Voigt, T. Tørring, A. Rotaru, M. F. Jacobsen, J. B. Ravnsbaek, R. Subramani, W. Mamdouh, J. Kjems, A. Mokhir, F. Besenbacher, and K. V. Gothelf, *Nature nanotechnology*, 2010, **5**, 200-3.
8. A. V. Pinheiro, D. Han, W. M. Shih, and H. Yan, *Nature Nanotechnology*, 2011, **6**, 763-772.
9. C. Martin, *Nature materials*, 2012, **11**, 95.
10.   L. J. Steinbock, O. Otto, C. Chimerel, J. Gornall, and U. F. Keyser, *Nano Letters*, 2010, **10**, 2493-2497.
11.   L. J. Steinbock, A. Lucas, O. Otto, and U. F. Keyser, *Electrophoresis*, 2012, **In press**.

ON THE DEVELOPMENT OF NEW METHODS FOR ION CHANNEL STRUCTURE-FUNCTION MEASUREMENT

J.W.F. Robertson, V. Silin, and J.J. Kasianowicz

NIST, Physical Measurement Laboratory, Semiconductor and Dimensional Metrology Division, CMOS Reliability and Advanced Devices Group, Gaithersburg, MD 20899-1070

1 INTRODUCTION

One of the major roadblocks to understanding cell function, and the development of therapeutic agents against disease, is the relative paucity of tools for determining integral membrane protein (IMP) structures.[1] Despite the extensive efforts applied to the problem, there are only ca. 300 solved structures of IMPs.[2] Unfortunately, of those with solved structures, many are the same protein from different species. Considering the medical and technological importance of IMPs, it is clear that new technologies need to be developed to address this metrology issue.

There are four major techniques for obtaining high resolution structures of proteins: x-ray crystallography,[2] electron microscopy,[3] NMR spectroscopy,[4,5] and EPR spectroscopy.[5,6-8] Each of these measurement modalities has its own advantages and limitations. For example, when they can be obtained, x-ray crystal structures provide much greater detail (atom level) well beyond the resolution of electron microscopy. However, the method does not verify whether the structure is that of a fully functional molecule. This is a significant concern, because the environment required for stable functional IMPs (a lipid bilayer membrane) is a stark contrast to that required to isolate the protein from the membrane and then to coax it to crystallize. NMR can provide estimates of molecular motion at atomic resolution, but the method is currently limited to the study of relatively small proteins (< 32 kg/mol) or protein fragments confined to small volumes or on surfaces.[9] The electron spin analog of NMR, EPR is used to elucidate membrane protein topology (e.g., the local structural motif of the protein: α-helix or β-sheet) and dynamics (e.g., whether a spin-containing amino acid side chain is in the membrane or aqueous phase).

While the contributions of these techniques to our present understanding of cell biology cannot be overstated, new technologies need to be developed that address corollary information and that can independently provide structural information and simultaneously provide evidence that the proteins actually work (there is essentially no value in knowing the structure of a non-functioning protein). Here, we discuss several such techniques, which are based upon the fabrication of a biomimetic membrane on a smooth metal electrode. The interface, a tethered bilayer lipid membrane (tBLM), is based upon self-assembly of lipid-like compounds to the electrode through thiol-chemistry with a spacer segment of a polymer.[10]   More recent approaches have used poly(ethylene glycol) as the spacer.[10-14]   Figure 1 (*top*) illustrates the membrane system in our experiments, which enables the use of a suite of surface-sensitive measurement modalities, including, but not limited to, electrochemistry (to test the protein's function) and a host of spectroscopic techniques to probe the structure of the molecule. A central goal of this system is to tailor the surface coverage of a given protein for the analytical method of choice.

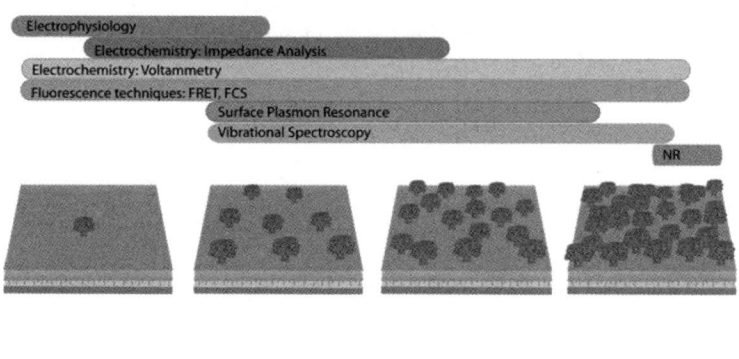

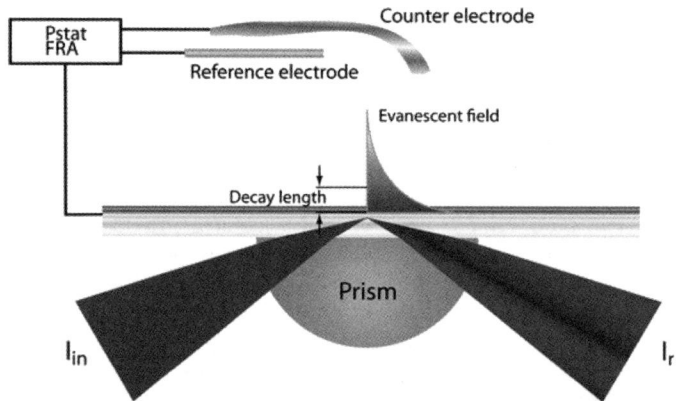

Figure 1 (*top*) Membrane constructs and methods used to study membrane proteins. A wide range of protein surface coverage can be probed using multimodal measurements that provide both functional and structural information about the proteins. (*bottom*) The ability to combine several methods (e.g., surface plasmon resonance spectroscopy to detect and characterize adsorbed proteins on the surface, and electrochemical impedance spectroscopy to test for ion channel formation) provides an estimate for the percentage of protein on the surface that functions properly.

2 METHODS AND RESULTS

For the study of bacterial pore forming toxins, one approach for optimizing the surface coverage of proteins is the simultaneous measurement of the resonance angle via surface plasmon resonance (SPR)[15] and the transmembrane resistance via electrochemical impedance spectroscopy (EIS) as depicted in Fig. 1 (*bottom*). The advantage of combining these measurements is that they sample different aspects of the protein's behavior. Surface plasmons can exist at a metal surface, are highly confined near the interface, and are highly sensitive for the dielectric constant of a thin film at the metal surface. The SPR signal is estimated from the minimum of reflected light at the angle of incidence when x component of the incident light wave vector that of surface plasmons. When the optical properties of the interface change (e.g., due to protein adsorption) the resonance angle shifts. With appropriate assumptions about the optical constants of the surface and proteins, a measurement of these shifts is used to estimate the protein surface concentration.[15] Conversely, EIS monitors changes in the ionic conductivity of the interfaces by applying a small AC potential wave and recording the current and phase shift over a wide range of frequencies, typically ~ 50 kHz to ~ 10 mHz. The data are subsequently analyzed by fitting to an idealized equivalent circuit model made of simple resistors and capacitors. For pore-forming toxins and other ion channels, the technique is highly sensitive with detection limits approaching the single molecule level,[16] and provides a wide dynamic range to observe channel formation over many orders of magnitude of surface coverage (Figure 1 (*top*)).

Recently published work on channel formation by the pore-forming toxin *Staphylococcus aureus* α-hemolysin[17] demonstrated the power of combining EIS and other measurement modalities (e.g., neutron reflectometry). Preliminary results for channel formation by the 63 kg/mol fragment (PA$_{63}$) of protective antigen secreted by *Bacillus anthracis* provide a good example of the synergy of simultaneous SPR and EIS measurements. A typical SPR measurement shows that upon addition of the PA$_{63}$, the SPR signal suggests that most of the protein adsorbs relatively quickly at the interface, but a slower adsorption process occurs as well (Fig. 2). The corresponding EIS signal (Fig. 3) initially shows a high membrane resistance (*red*) upon injection of protein into the bulk solution this resistivity drops smoothly as a function of time (*blue, green, purple and black*) with a corresponding shift in the frequency of the minimum phase shift.   Taken together, these results demonstrate that some of the protein adsorbed to the membrane surface penetrates the membrane and forms ionic channels.[18-20] However, a comparison of the mass of adsorbed protein (estimated from the SPR signal) to the surface coverage of ion channels (estimated from EIS and the known conductance of single PA$_{63}$ ion channels,[21,18] suggests that much of the adsorbed protein does not form channels. Although the SPR data indicate that the protein is absorbed at significant surface coverage (Fig.2), only a small fraction insert their β-barrel into the membrane are anchored to the surface, suggesting that part of the protein can denature and lose the ability to form channels. The added advantage here is that ion channels can be tested for function prior to, or simultaneously to, application of structural techniques (*see below*), which has not been demonstrated with any other technology.[17]

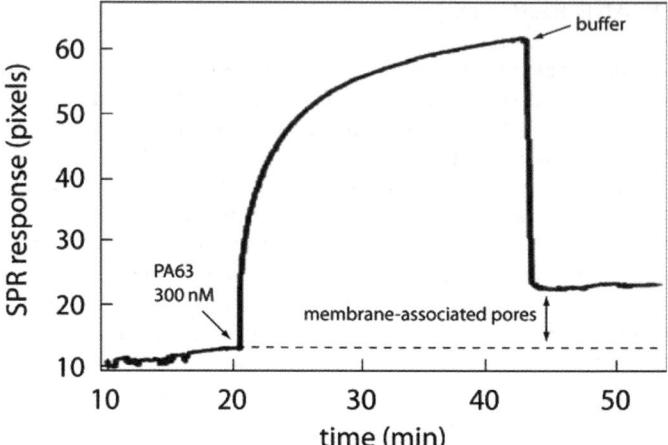

Figure 2. Time dependent surface plasmon resonance measurement of a tethered bilayer membrane after adding the pore-forming toxin, *B. anthracis* PA_{63} to the electrolyte solution bathing the membrane (0.1 M KCl, pH 6.6).

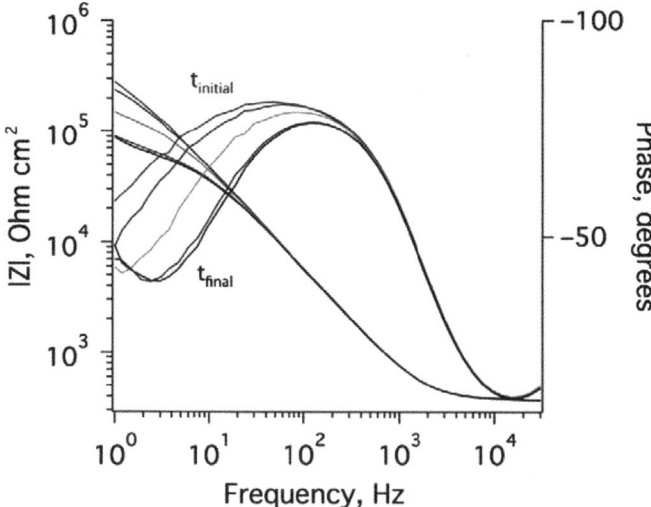

Figure 3. Kinetics of *B. anthracis* PA_{63} channel formation in tethered bilayer membranes determined with electrochemical impedance spectroscopy. Channel formation reduces the low frequency modulus and phase angle by reducing the membrane resistance.

Our initial neutron reflectometry (NR) experiments (*data not shown*) were hampered by low surface coverage of PA_{63} channels, but strongly suggest that this approach will ultimately succeed as it did for the α-hemolysin channel.[17] As with x-ray reflectometry, the neutron intensity decreases dramatically as a function of momentum transfer. Data were collected for both the membrane and the protein-infused membrane under three different solvent contrasts -- neutrons interact with the nucleus of atoms and protons and deuterium offer the largest difference between two isotopes of any element that occurs in biological systems.[22] Features in the data are dominated by the interference pattern from neutrons reflected off the various interfaces in the system. To develop a picture of the interface, layer-by-layer models must be constructed and tested against these data. These preliminary results suggest that the surface coverage is currently only between 1 % and 3 %, which does not provide sufficient detail to make significant claims about the structure of this pore-forming toxin.

3 CONCLUSION

Biomimetic interfaces can be constructed on atomically smooth electrode surfaces to provide a measurement platform that allows both high-information structural experiments such as NR, or vibrational spectroscopy as well as allow functional measurements of the protein as well. The true advantage of the robust tethered bilayer [23] is the ability to make many different measurements over a long period of time on the exact same interface. No other technology affords this ability.

4 ACKNOWLEDGMENTS AND NOTES

Supported in part by the NIST Office of Law Enforcement Standards. Certain methods are mentioned in the manuscript. This in no way represents an endorsement by NIST.

References

1. D. Rees, G. Chang, and R. Spencer, *J. Biol. Chem.*, 2000, **275**, 713–716.
2. S. D. White, Ed. *Membrane Proteins of Known 3D Structure* at http://blanco.biomol.uci.edu/mpstruc/listAll/list
3. R. Henderson and P. Unwin, *Nature*, 1975, **257**, 28–32.
4. *Membrane Proteins of Known Structure by NMR.* http://www.drorlist.com/nmr/MPNMR.html
5. U. A. Hellmich and C. Glaubitz, *Biological Chemistry*, 2009, **390**, 815–834.
6. B. Honig and W. Hubbell, *Proc. Natl. Acad. Sci. (USA)*, 1984, **81**, 5412–5416.
7. W. L. Hubbell, D. S. Cafiso, and C. Altenbach, *Nat. Struct Biol.*, 2000, **7**, 735–739.
8. C. Altenbach, T. Marti, H. Khorana, and W. Hubbell, *Science*, 1990, **248**, 1088–1092.
9. K. Wüthrich, *NMR of proteins and nucleic acids*, Wiley-Interscience, 1986.
10. B. Cornell, V. Braach-Maksvytis, L. King, P. Osman, B. Raguse, L. Wieczorek, and R. Pace, *Nature*, 1997, **387**, 580–583.
11. A. T. A. Jenkins, R. Bushby, N. Boden, S. Evans, P. Knowles, Q. Liu, R. Miles, and S. Ogier, *Langmuir*, 1998, **14**, 4675–4678.
12. L. He, J. W. F. Robertson, J. Li, I. Karcher, S. Schiller, W. Knoll, and R. L. C. Naumann, *Langmuir*, 2005, **21**, 11666–11672.
13. F. Giess, M. G. Friedrich, J. Heberle, R. L. C. Naumann, and W. Knoll, *Biophys J*, 2004, **87**, 3213–3220.
14. V. Atanasov, P. P. Atanasova, I. K. Vockenroth, N. Knorr, and I. Köper, *Bioconjugate Chem.*, 2006, **17**, 631–637.
15. W. Knoll, *Annu. Rev. Phys. Chem.*, 1998, **49**, 569–638.
16. I. K. Vockenroth, P. P. Atanasova, A. T. A. Jenkins, and I. Köper, *Langmuir*, 2008, **24**, 496–502.

17. D. J. McGillivray, G. Valincius, F. Heinrich, J. W. F. Robertson, D. J. Vanderah, W. Febo-Ayala, I. Ignatjev, M. Losche, and J. J. Kasianowicz, *Biophys. J.*, 2009, **96**, 1547–1553.
18. B. J. Nablo, K. M. Halverson, J. W. F. Robertson, T. L. Nguyen, R. G. Panchal, R. Gussio, S. Bavari, O. V. Krasilnikov, and J. J. Kasianowicz, *Biophys. J.*, 2008, **95**, 1157–1164.
19. R. Blaustein, T. Koehler, R. J. Collier, and A. Finkelstein, *Proc. Natl. Acad. Sci. (USA)*, 1989, **86**, 2209–2213.
20. R. Blaustein and A. Finkelstein, *J. Gen. Physiol.*, 1990, **96**, 905–919.
21. K. M. Halverson, R. G. Panchal, T. L. Nguyen, R. Gussio, S. F. Little, M. Misakian, S. Bavari, and J. J. Kasianowicz, *J. Biol. Chem.*, 2005, **280**, 34056–34062.
22. J. Fitter, T. Gutberlet, and J. Katsaras, *Neutron scattering in biology*, Springer Verlag, 2006.
23. I. K. Vockenroth, C. Ohm, J. W. F. Robertson, D. J. McGillivray, M. Losche, and I. Köper, *Biointerphases*, 2008, **3**, FA68.

IMPROVED ALGORITHMS FOR NANOPORE SIGNAL PROCESSING

N. Arjmandi[1,2], W. Van Roy[1], L. Lagae[1,2], G. Borghs[1,2]

[1] IMEC, Kapeldreef 75, 3001 Leuven, Belgium
[2] Department of Physics and Astronomy, KU Leuven, Celestijnenlaan, 200D, 3001 Leuven, Belgium

1 INTRODUCTION

Nanopores are promising devices for detection and characterization of nanometer-sized analytes suspended in a liquid; these include nanoparticles [2], DNA [3], RNA [4], viruses [2] and proteins [5]. These devices are basically a nanometer-sized pore that connects two microfluidic chambers. Translocation of analytes through the pore introduces a temporal change in its resistance that can be recorded as a spike in the ionic current that passes through the pore or a voltage spike. All applications of nanopore devices are based on detection of translocation spikes in the recorded signal and extraction of amplitude, duration and their rate of occurrence [2-5]. The recorded signal is usually considerably noisy [6], with a significant baseline drift [6] and more than hundreds of translocation spikes that may vary in shape and size [2]. Thus, incorporation of suitable signal processing algorithms is necessary for correct and fast detection of all the translocation spikes and accurate measurement of their amplitude and duration. Nanopore devices are subjected to intense research and significant improvements; however, there are only a few reported investigations on processing the output signals of these devices [7-14]; while these methods are determinant of the technique's precision, accuracy and applicability.

Nanopore signal processing generally consists of baseline removal, denoising, spike detection, extraction of spikes' amplitude and duration, and classification of spikes. Here we present an improved method for baseline removing, an optimized algorithm for denoising the nanopore signals, a novel method for spike detection that detects all the translocation spikes more correctly, and an improved and physically meaningful algorithm for measuring the duration and amplitude of the translocation spikes [1].

2 METHOD AND RESULTS

2.1 Experimental Results
The proposed software has been used to process ionic current signals of a solid-state nanopore (fig. 1) using a range of different nanoparticles. The obtained results are compared with the results of the conventional software (Clampfit, MDS Analytical Technologies) in addition to results of a commercially available dynamic light scattering (DLS) system. In order to fabricate the nanopore, silicon on insulator wafers were coated by nitride. Electron beam lithography [15] was used for patterning a square on the front

side and KOH was used for etching a pyramidal pit into the silicon on the front side. Then, a pyramidal pit was etched on the back side to open the nanopore's end, and the nanopore was packaged between two microfluidic chambers as described elsewhere [16]. Citrated gold nanoparticles (Ted Pella Inc.) were diluted ten times in 300mM KCl solutions, and put on the front side of the nanopore device. 120 mV was applied between the two sides of the nanopore and the ionic current recorded by an Axopatch 200B patch clamp amplifier. A MiniDigi (Molecular Devices) was used for analog to digital conversion and pCLAMP 10 (Molecular Devices) was used for data acquisition.

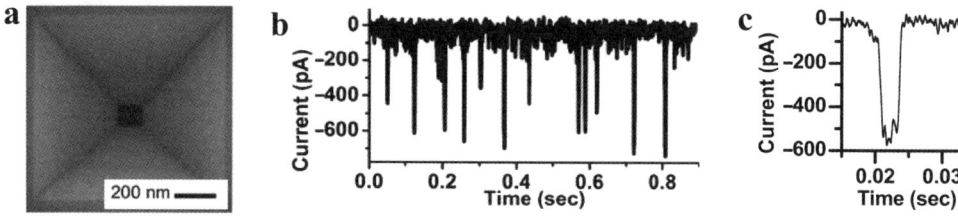

Figure 1 *The nanopore experiment. a) Top view of a 120 nm wide pyramidal nanopore. b) A recorded ionic signal resulting from a mixture of particles as mentioned in the text. c) Close-up view of a typical translocation spike.*

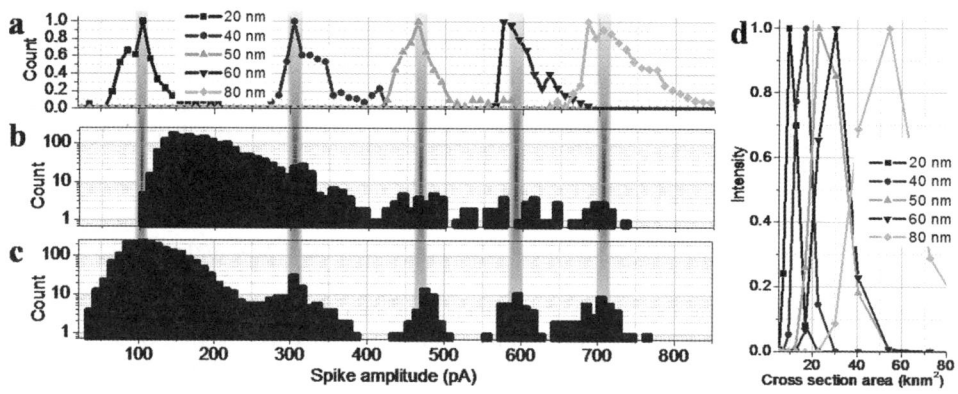

Figure 2 *Comparing the conventional and the proposed methods. a) Normalized histogram of spike's amplitudes obtained by the proposed method from separate nanopore measurements using 20, 40, 50, 60 and 80 nm gold nanoparticles. b) Histogram resulting from the conventional thresholding method from a mixture of the aforementioned particles. c) Histogram obtained from the proposed method from the same signal. d) Separate DLS measurements results of the particles.*

Different types of particles result in different amplitudes (fig. 2.a, d). We have obtained narrower distributions, more precise and better-separated populations, using the proposed software comparing to the conventional methods. To present these properties in a practical experiment, we have used a mixture of 20, 40, 50, 60, and 80 nm nanoparticles and used different methods to process the recorded signal. Conventional methods do not result in well separate populations at the expected points (fig. 2.b), while the proposed software results in more accurate and separable populations (fig. 2.c). The origin of such an improvement and the details of the proposed methods are described hereunder.

2.2 Baseline Removing

Although, the proposed software measures the amplitude and duration of the translocation spikes independent of signal's baseline, it first removes the baseline to enhance the visualization of the signals in addition to extraction of the equivalent circuit parameters of the device. The baseline drift mainly originates from the membrane as a lossy capacitance [6]. The most widely used method for baseline removing is the so called moving window (MW), which is averaging a number of data points around each translocation spike [2,8,14]. These algorithms are prone to significant inaccuracies in baseline detection in some frequently occurring conditions (fig. 3). Considerating the membrane as a lossy capacitor [6], the nanopore as a resistor, bulk liquid as a resistor, and electrodes as electric double layer capacitors in parallel with tunneling resistors, results in a summation of infinite number of exponentials for the step response of the nanopore. However, we found sufficiently precise to consider a double exponential. Hereby, the baseline can be found by fitting this summation of exponentials to the upper (or lower) envelope (as appropriate) of the signal.

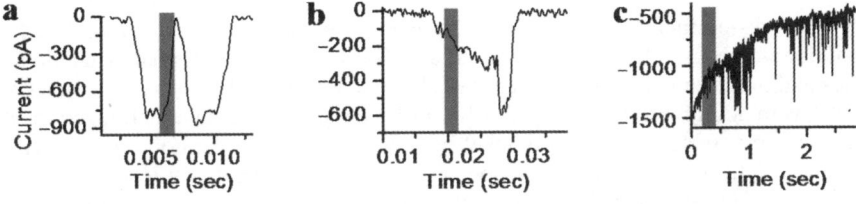

Figure 3 *Some deficiencies of the moving window (MW) baseline detection. a) When there are two adjacent spikes, MW detects the first spikes as the others baseline. Here although the baseline is 0 pA, the MW has detected -381 pA. b) When spikes have a tail. Here although the baseline is 0 pA, MW has detected -175 pA. c) Since is not possible to reduce the window too much. MW's error increases when the baseline drifts too fast.*

2.3 Denoising

Low-pass filtering is generally used for denoising the nanopore signals [2, 6, 17, 18] and introduces a trade-off between signal to noise ratio and measurement bandwidth. Due to the fact that the translocation spikes are pulse shaped [2-14] and contain a broad range of frequencies, low-pass filters can only be used at very high cut off frequencies to remove the digitization noise. In addition to using a high cut off frequency, proper choice of filter is necessary to preserve the signal. To choose the best filter, we have examined Bessel 8-pole, Gaussian, Butterworth 8-pole, Boxcar, Chebyshev 8-pole, RC 8-pole and RC 1-pole low pass filters. Although all of these filters can remove the digitization noise, the least deformation and phase shift in the signal achieved with the Gaussian filter.

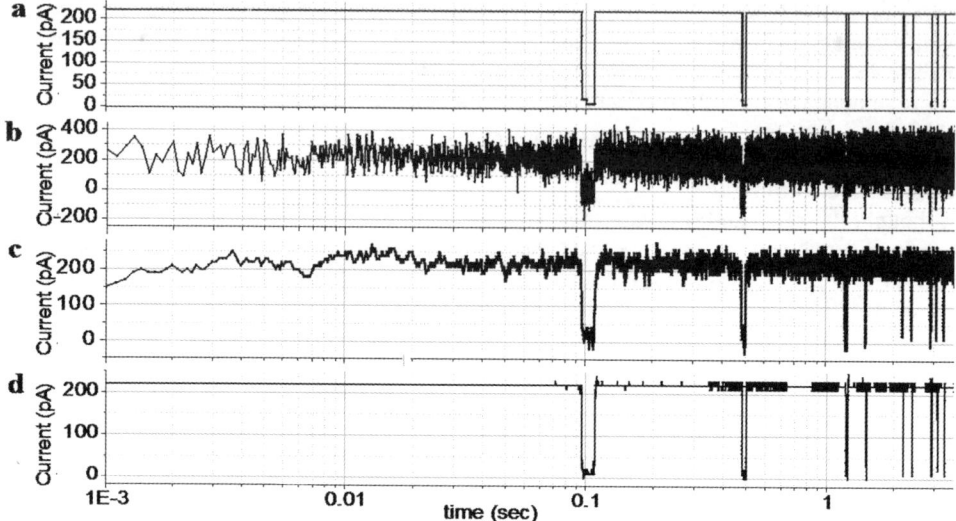

Figure 4 *Simulated signals (time axis is scaled logarithmically). a) 10 typical spikes. b) The simulated noise added to the simulated spikes (10 kHz sampling rate). c) Low-pass filtered of the signal in b. d) The signal in c after denoising by Bior3.9.*

A proper denoising increases the detection range of the nanopore by enabling the detection of smaller analytes in a larger nanopore, in addition to detection of smaller features of the analytes and increased precision. Jagtiani et al. have described denoising of Coulter signals by wavelets in addition to the low-pass filter [19]. These authors have optimized the choice of wavelet and wavelet thresholding level by a cross validation method [20, 21]. However this method optimizes the general shape of the denoised signal, while we are interested mostly in its amplitude and duration; furthermore, it is a probabilistic approach and does not result in certain results. Thornton et al. [7, 9, 13] have introduced wavelet denoising for nanopore signals without optimizing it. However, wavelet denoising is a strong function of type of wavelet, threshold level and signal's shape. In addition, optimization of the denoising for accurate extraction of amplitude and duration is needed. Thus, we have examined 54 different wavelets at 15 thresholding levels to choose the best denoising configuration. For this purpose, all the forces that are acting on a nanoparticle in a 120 nm nanopore have been calculated and a Monte Carlo simulation was used to calculate the nanoparticle's trajectory and the resulting spikes (fig. 4.a). The shapes of these spikes are in agreement with the experimental measurements (fig. 1.c). Ten translocation spikes were simulated, noise with the same power spectrum of the experiment is generated and added to the simulated signal (fig. 4.b); and then it was low-pass filtered as mentioned earlier. Different denoising algorithms and configurations were applied to this signal and the integral of root square error (IRSE) of the denoised signal in comparison to the originally simulated noise-less signal was calculated (fig. 5.a). About 9 different denoising configurations, which are resulting in the smallest IRSEs, were found. To choose the best of them relying on IRSE is not reliable and optimal; thus, the amplitude and duration of the denoised spikes measured and the correlation coefficients of these parameters with the originally simulated noise-less spikes have been calculated (fig. 5.b). Denoising by Bior3.9, results in the highest correlation between the amplitudes of the noise-less signal and the denoised signal. Denoising by rBio1.3 is relatively better for duration measurement. Coif2 is the best for measuring the amplitude and duration at the same time.

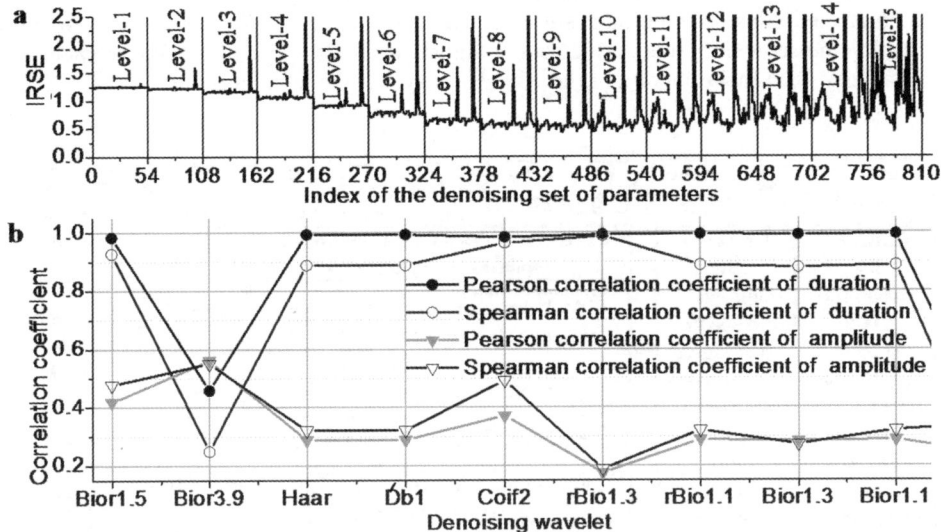

Figure 5 *Comparing the denoising algorithms. a) IRSE of the signals that are denoised by 810 different methods and conditions. These methods are consisting of 54 different wavelets in 15 different levels. The used wavelets are: Demy, Db1 to 10, Sym2 to 8, Coif1 to 5, Bior1.1, 1.3, 1.5, 2.2, 2.4, 2.6, 2.8, 3.3, 3.5, 3.7, 3.9, 4.4, 5.5, and 6.8, rBio1.1, 1.3, 1.5, 2.2, 2.4, 2.6, 2.8, 3.3, 3.5, 3.7, 3.9, 4.4, 5.5 and 6.8. b) Comparison of the correlation coefficients of the 9 wavelets that have the lowest IRSE.*

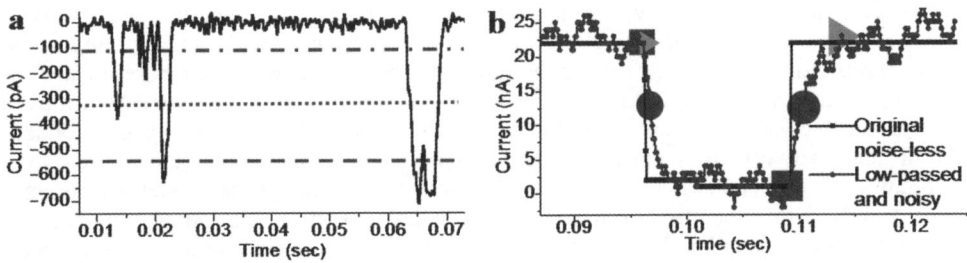

Figure 6 *Spike analysis. a) Possible threshold choices. b) Starting and ending points as determined by the proposed algorithm (■) and thresholding (baseline: ▶, FWHM: ●).*

2.4 Spike Detection

The current method for spike detection is the so called thresholding [2-14, 22]. In this method, a level about 5 times the noise level away from the base line is chosen; and when the signal crosses this level, the algorithm reports a spike; fluctuations which are not crossing the threshold are left undetected (Fig. 6.a). Using this approach it is difficult to detect both small spikes and closely separated double peaks.

To solve this problem we have developed a novel algorithm, which detects any fluctuation in the signal which is physically meaningful – i.e. it is not probable to be originating from the electrical noise. This algorithm can detect the fluctuations within a spike, in addition to differentiation of adjacent spikes. To do so, it extracts all the local minima and maxima of the signal and considers the adjacent extrema with amplitude difference more than five times the noise level. Then, a selection algorithm finds the spikes and extracts their parameters. The block-diagram of the whole software is depicted in figure 7.

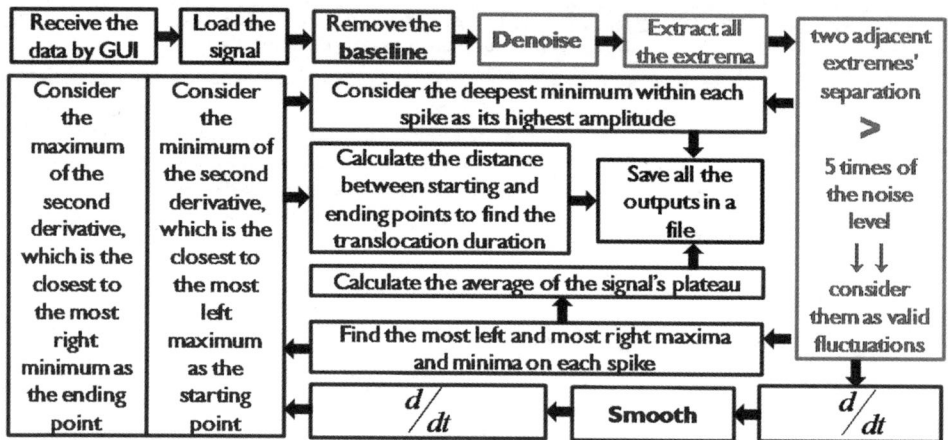

Figure 7 *Block diagram of the proposed software.*

2.5 Amplitude and Duration Determination

Previously, there were a few slightly different approaches for measuring the spike's duration, all based on thresholding; the distance between the two adjacent crossing points of the threshold, the distance between the point that the signals leaves the baseline and the point that it returns to the baseline [6, 23, 24, 25], or considering the full width half max of the spike to reduce the effects of the limited bandwidth [2]. Pedone et al. have introduced the distance between the point at which the signal leaves the baseline and the last local minimum of the spike as the spike's duration [8]. Although such definitions may considerably reduce the effect of bandwidth, they have no physical meaning and are prone to noise within the spike. Thus, a new definition for the spike's duration is introduced, which is least affected by the measurement bandwidth and noise, and most importantly has a physical meaning. This definition considers the time that the center of nanoparticle enters the sensing zone as the start of the spike and the time that its center leaves the sensing zone as the ending point. A simple analytical calculation shows that these are the times at which the second derivation of the signal reaches its minimum and maximum. Since the measurement bandwidth only changes the rise time and fall time of the signal and maybe the peak amplitude of the signal, it is not affecting the spike's duration that has measured by this method (fig. 6.b).

Conventional algorithms are considering the distance between the average of the spike's plateau or its highest point, to the baseline as the spike's amplitude. In addition to these amplitudes, the proposed software also calculated the distance between the highest maximum and minimum within a spike as its amplitude. The later definition of the amplitude is not affected by the baseline drift and results in narrower distributions.

References

1   The developed software and extensive supplementary information will be made available to the interested reader upon request.
2   J. L.Fraikin et al., Nature Nanotechnology, **6**, 2011, 308
3   B. Murali Venkatesan, R. Bashir, Nature Nanotechnology, **6**, 2011, 615
4   Y. Wang et al., Nature Nanotechnology, **6**, 2011, 668
5   E. C. Yusko et al., Nature Nanotechnology, **6**, 2011, 253
6   R. M. M. Smeets, U. F. Keyser, N. H. Dekker, and C. Dekker, PNAS, **105**, 2008, 417
7   A. Spanias et al., Signal Processing Applications for Public…, 2007, 1
8   D. Pedone, M. Firnkes, U. Rant, Anal Chem, **81**, 2009, 9689
9   B. Konnanath et al., ICANN 2009, 2009, 265
10 P. Joshi et al., 2009 ASME IMECE, Florida, 2009, 11428
11 T. Mathew et al., Advanced Smart Materials and Smart Structures, Boston, 2009
12 Z. Cai et al., Biomedical Engineering and Informatics, 2010, **4**, 1373
13 P. Sattigeri et al., Communications, Control and Signal Processing, 2010, 1
14 J. A. Billo, W. Asghar, S. M. Iqbal , Proc. SPIE 8031, 2011, 80312T
15 N. Arjmandi, L. Lagae, G. Borghs , J. Vac. Sci. Technol. B **27**, 2009, 1915
16 N. Arjmandi et al., Microfluidics and Nanofluidics, **12,** 2011, 17
17 V. Tabard-Cossa et al, Nanotechnology, **18**, 2007, 305505
18 J. D. Uram, K. Ke, M. Mayer, ASC Nano, **2**, 2008, 857
19 A. V. Jagtiani, R. Sawant, J. Carletta, J. Zhe, Meas. Sci. Technology, **19**, 2008, 065102
20 G. P. Nason, J R Stat Soc Series B, **58**, 1996, 463
21 L. Pasti et al., Chemometr Intell Lab Sys, **48**, 1999, 21
22 L.J. Steinbock, G. Stober, U.F. Keyser, Biosensors and Bioelectronics, **24**, 2009, 2423
23 S. M. Iqbal, D. Akin, R. Bashir, Nature Nanotechnology, 2007, **2**, 243
24 L. T. Sexton et al., Am. Chem. Soc., **129**, 2007, 13144
25 J. B. Heng et al., Biophys. J., **87**, 2004, 2905

A VARIABLE CROSS-SECTION PORE FOR SCREENING CELLS FOR SPECIFIC MARKERS

K. Balakrishnan[1], M. Chapman[2], A. Kesavaraju[3], L. Sohn[1]

[1] Department of Mechanical Engineering, University of California, Berkeley, CA, 94720, USA
[2] Biophysics Graduate Group, University of California, Berkeley, CA, 94720, USA
[3] Departmnet of Bioengineering, University of California, Berkeley, CA, 94720, USA

1 INTRODUCTION

Nanopores have emerged as a versatile tool for performing highly sensitive single-molecule measurements to probe the properties of proteins and nucleic acids.[1,2] Recent work[3] has shown that interactions within a nanopore (due to pore functionalization with metals, oxides, or organic species) can result in slower translocation rates, thereby realizing "smart" nanopore sensors that provide insight into the properties of the particles that are transiting the pore. At the same time, while greatly enhancing its utility, functionalizing a pore with only one type of species limits the broader applicability of pores for biosensing.

We have addressed this limitation by developing a variable cross-section pore that creates unique electronic signatures for reliable detection and automated data analysis[4]. By using common lithography techniques to define a single pore into separate sections, we are able to use resistive-pulse sensing[5] to determine precisely when a biological cell passes through a given pore segment. By having the capability to functionalize multiple segments of the pore with different antibodies that can transiently interact with complementary antigens on the cell's surface and slow the cell down, we can potentially determine the presence of specific markers on the cell when we compare transit times of each section. As will be described below, we have achieved proof-of-principle of a single-marker with our newly designed pore. Overall, our variable cross-section pore offers specific advantages: 1) the ability to functionalize each portion of the pore specifically with a different antibody that corresponds to different cell-surface receptors, which in turn enables label-free multimarker detection in a single run; and 2) a unique electronic signature that allows for both an accelerated real-time analysis and an additional level of precision to testing. This is particularly critical for clinical diagnostics where accuracy and reliability of results are crucial for healthcare professionals upon which to act.

2 METHOD AND RESULTS

2.1 Variable Cross-section Pores

Utilizing the technique of resistive-pulse sensing, particles (e.g. cells) transiting a pore can be measured for size and transit time. The current across the pore changes due to the passage of particles whose resistivity is greater than that of the solution in which they are suspended. This current change can be understood by examining the cross-sectional slices of the pore through which the cell travels[6], as shown in gray in Figure 1. Assuming a non-conducting particle and pore with radius $r_{particle}$ and r_{pore}, respectively, the resistance of a disk of thickness dx is

$$dR = \frac{\rho_{fluid}\, dx}{\pi\,(r_{pore}^2 - r^2)}$$

Eq. 1

where ρ_{fluid} corresponds to the fluid resistivity. Because

$$r^2 = r_{particle}^2 - x^2 \qquad if\; x^2 \leq r_{particle}^2$$
$$r^2 = 0 \qquad if\; x^2 > r_{particle}^2$$

Eq. 2

where r is the height of the slice and x is the distance of the slice from the center of the particle along the axis of the pore, we can express the increase in resistance due to the presence of a particle as,

$$\Delta R = \frac{\rho}{\pi}\int_{-r_{particle}}^{r_{particle}} \frac{dx}{r_{pore}^2 - r_{particle}^2 + x^2} - \frac{\rho}{\pi}\int_{-r_{particle}}^{r_{particle}} \frac{dx}{r_{pore}^2}$$

Eq. 3

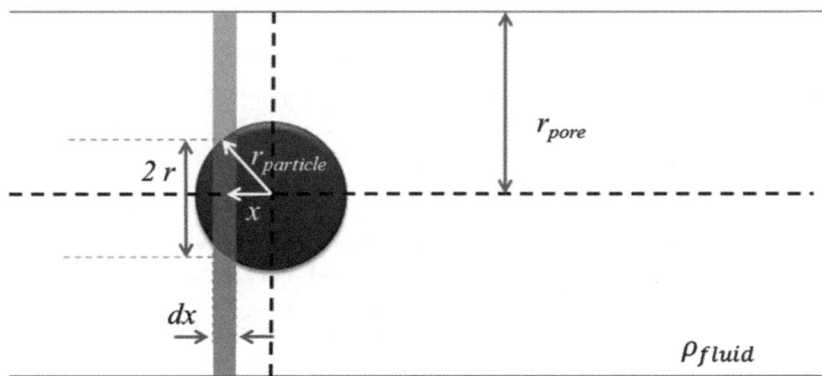

Figure 1: Schematic showing a spherical non-conducting particle of radius $r_{particle}$ in a pore of radius r_{pore} and filled with conducting fluid.

Thus, if a pore has a constant cross-sectional area, so too will the change in resistance caused by the particle transiting the pore be constant (Eq. 3). Likewise, if the cross-sectional area of the pore changes, then the resistance (Eq. 3) will also change as the particle transits the pore. By utilizing the fact that a resistance measurement across the pore at any given time is dependent on r_{pore} of each cross-sectional slice of width dx at that particular time, one can specifically tailor the shape of the pore to provide a desired current signal measurement.

Based on this analysis, we demonstrate that by inserting a series of "notches" into a pore (Figures 2a and b), we can precisely determine when a cell has traversed a given region of that pore. As shown in Figure 2c, the current (resistance) modulates as the cell travels through the pore. This modulation creates electronic detection "markers" in a current vs. time measurement that can, as will be shown below, increase the reliability and ease of data processing.

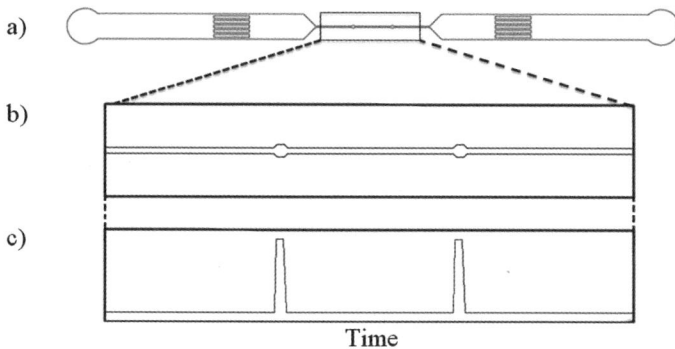

Figure 2: a) *Schematic of a "notch" pore device with two reservoirs and two sets of filters. b) A magnified view of the pore, containing two notches.   c) The corresponding current vs. time measurement as a particle transits the pore and through each notch.*

Variable sized pores are created using standard soft-lithography techniques. Negative-relief masters, containing pores, filters, and reservoirs (Figure 2a), are fabricated on polished silicon wafers using conventional photolithography. The cross-sectional area of the pores are 25 μm x 25 μm for easy passage of cells one-by-one, and the pore length is 2000 μm for clear transit-time measurements. Degassed PDMS (Sylgard 184, 10:1 pre-polymer : curing agent) is dispensed onto the negative-relief master and cured for an hour at 80°C. PDMS slabs with the embedded channels and pores are cut from the masters and cored for inlet and outlet holes. The slabs are then sealed to glass substrates which already have Pt electrodes lithographically patterned onto them. Functionalizing of the pore with antibody is accomplished using methods described in previous work[5].

2.2 Unique Electronic Signatures

To measure the current across the pore, we utilize a four-point measurement with a constant applied AC voltage (typically 0.2–0.4 V), as previously published.[5,7,8,9] Data are sampled at 50 kHz, recorded, and analyzed using custom-written software in LabVIEW. A constant pressure of 3 psi is used to drive cells across the filters, reservoirs, and pore.

HCT116 colorectal cancer cells suspended in McCoy's 5a medium with 10% FBS were screened with pores containing several notches of variable spacing (Figure 3a). As shown in Figure 3b, the raw data of a typical measurement is difficult to interpret. However, after low pass filtering, we find that a pulse, corresponding to a cell traveling through the pore, becomes apparent, and by focusing in on the region of interest (Figure 3c), we are able to resolve further the individual cell transit times corresponding to each region of the pore (indicated as 1-8) and to determine precisely when the cell traversed a specific section.

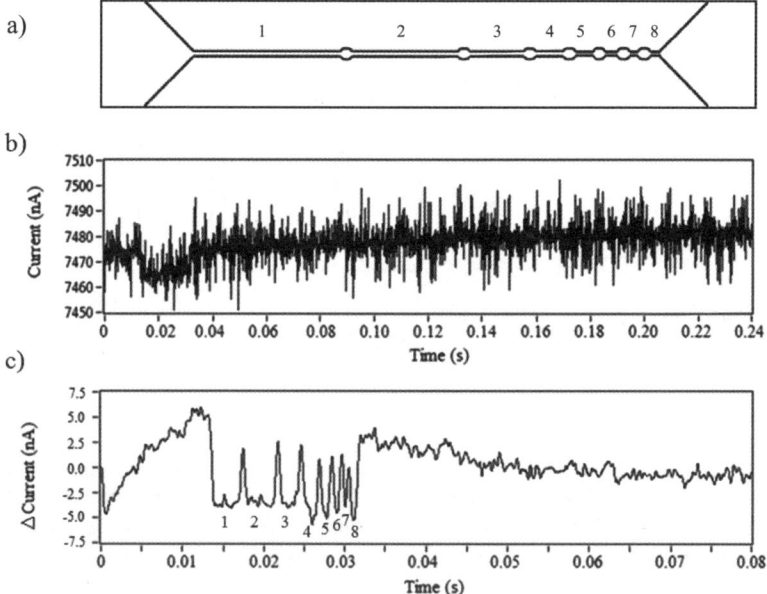

Figure 3: *Measuring HCT116 colorectal cancer cells with a pore with several notches with variable spacing. a) Schematic of the pore used. b) Shows raw data and c) shows a magnified view of the region of interest after a low pass filter was applied.*

Thus, an additional advantage of inserting notches into the pore is that we are able to not only identify exact transit-time locations, but also develop unique electronic signatures that can readily improve data analysis. While a single transit-time pulse can be hidden by noise, signal-to-noise ratios become less important when examining a unique signature. Thus, as in Figure 3, a pulse with several pulses within it is easily identifiable and can allow for quicker and more accurate data processing.

2.3 Transit-Time Normalization and Surface-Marker Screening

One can utilize notched pores when screening for a single cell-surface marker by comparing the transit time over a functionalized region with that of unfunctionalized or "blank" regions. Here, we show that by functionalizing only the section between two notches with a specific antibody (Figure 4a), we can standardize all our measurements. Transit times for a single cell passing through the functionalized section of the pore can be normalized with respect to the average of the two transit times, τ_1 and τ_3 (Figure 4b). In so doing, we remove any potential fluctuations in flow rate that could affect transit time measurements and analysis. Cells are positive for a specific marker if specific interactions between the marker and the functionalization species occur, retarding the cell as it travels through the middle region of the pore. This means that the normalized transit time value, τ_{norm} , will have a value larger than 1.

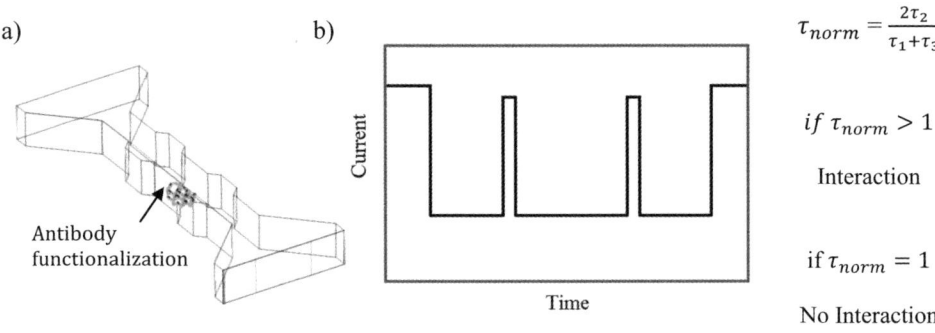

a)   b)

$$\tau_{norm} = \frac{2\tau_2}{\tau_1 + \tau_3}$$

if $\tau_{norm} > 1$

Interaction

Antibody functionalization

if $\tau_{norm} = 1$

No Interaction

Time

Figure 4: *A schematic describing how cells are screened for a particular marker when measured with a two-notched pore device. a) Drawing of a two-notched pore device with the middle region functionalized with antibody. b) The corresponding screening measurement made as a cell transits the pore. Pulse 1 and 3 correspond to the blank regions in the pore and have transit times τ_1 and τ_3. Pulse 2 corresponds to the functionalized region of the pore and has a transit time τ_2. $\tau_{norm} = 1$ if there is no interaction between the cell and the functionalized surface, and $\tau_{norm} > 1$ if there is interaction between the cell and the functionalized surface.*

2.4 Surface-Marker Screening Using Notched Pores

The notched pore device that we have just described in the previous section can be used to screen for specific cell-surface markers. Figure 5 shows data obtained when we screened HT29 colorectal adenocarcinoma cancer cells for EpCAM. The modulated current pulse is clearly visible. The transit times of the two blank regions, τ_1 and τ_3, are 5.48 ms and 5.58 ms respectively. The transit time of the functionalized (with anti-EpCAM antibody) region is 6.10 ms. The normalized transit time value is thus 1.10, indicating that the cell is EpCAM+.

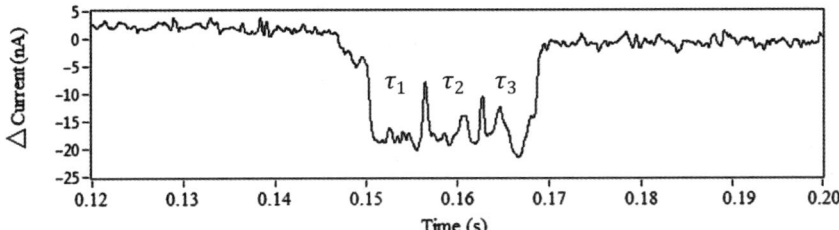

Figure 5: *Screening of HT29 colorectal adenocarcinoma cells for EpCAM using a 2-notched device whose central section has been functionalized with anti-EpCAM antibody. As shown in Figure 4, the two notches are equally spaced. For this particular pulse caused by a cell transiting the pore, the normalized transit time, τ_{norm} is 1.10. This value is larger than one, indicating that there is specific interaction occurring between the cell and the functionalized antibodies, i.e. the cell is EpCAM+.*

Because the introduction of notches facilitates the exact location and time extraction for cells transiting a pore, cell screening techniques using resistive pulse sensing can be greatly improved upon. The technology allows for the ability to functionalize each region of a pore differently, and detect the transit times of cells as they pass through each region of functionalization. This ultimately opens up a much broader applicability for resistive pulse sensors. Multiple antigen screening can be accomplished using a single pore, without the need for additional fabrication capabilities.

3 CONCLUSION

We have demonstrated a unique solid-state pore device that is capable of measuring multiple protein markers on a cell's surface. Our findings show that not only does our method increase the effective throughput by reducing the number of runs needed to characterize a particle in transit, but it also eliminates any issues with device-to-device variation that may be present with current solid-state pores. With the use of notched-pore devices, the sensing capabilities of several resistive-pulse measuring chips can be condensed into a single chip, offering obvious fabrication and cost advantages as well as reliability and consistency of testing.

References

1 H. Baley, *Current Opinion in Chemical Biology*, 2006, **10**, 628-637.
2 J. J. Kasianowicz, E. Brandin, D. Branton, D. W. Deamer, *Proc. Natl. Acad. Sci.*, 1996, **93**, 13770-13773.
3 M. Wanunu and A. Meller, *Nano Letters*, 2007, **7**, 1580-1585.
4 K. Balakrishnan, M. R. Chapman, A. Kesavaraju, G. Anwar, L. L. Sohn, in preparation.
5 A. Carbonaro, S. K. Mohanty, H. Huang, L.A. Godley, L. L. Sohn, *Lab Chip*, 2008, **8**, 1478-1785.
6 E. C. Gregg and K. D. Steidley, *Biophysics J.*, 1965, **5**, 393-405.
7 O. Saleh, L. L. Sohn, *Rev. Sci. Instrum.*, 2001, **72**, 4449-4451.
8 O. Saleh, L. L. Sohn, *Nano Letters*, 2003, **3**, 37-38.
9 O. Saleh, L. L. Sohn, *Proc Natl Acad Sci*, 2003, **100**, 820-824.

SALT AND VOLTAGE DEPENDENCE OF THE CONDUCTANCE BLOCKADE INDUCED BY TRANSLOCATION OF DNA AND RECA FILAMENTS THROUGH SOLID-STATE NANOPORES

Stefan W. Kowalczyk and Cees Dekker

Department of Bionanoscience, Kavli Institute of Nanoscience, Delft University of Technology, Lorentzweg 1, 2628 CJ, Delft, The Netherlands

1 INTRODUCTION

Over the course of the last decade, research on nanopores has grown into a mature field.[1-6] Amongst many other applications, solid-state nanopores – tiny holes in a thin membrane – can be used for rapid single-molecule high-throughput label-free detection of small samples of, for example, DNA[1] or DNA-protein[7] complexes. The interpretation of the current signal that one observes when DNA transverses a nanopore is a central point in all nanopore work. However, a complete and thorough understanding is still lacking.

Here, we present measurements of the conductance blockades (ΔG) for dsDNA translocations in a wide range of salt concentrations (20 mM – 1 M) and voltages (10 – 1000 mV). A clear trend of increasing ΔG with voltage is found for DNA, while ΔG is found to remain independent of voltage for RecA-DNA filament translocations. At low salt concentrations this leads to sign reversal of ΔG as a function of voltage for DNA translocations. We present several (mutually not exclusive) scenarios to explain the data. In particular, two scenarios stand out: in one scenario an increased access resistance due to the presence of the translocating molecule near the pore entrance leads to higher ΔG values at high voltages; in the other scenario a reduced number of counterions along the DNA leads to a lower current enhancement at high voltages.

2 METHODS

Our experimental setup for measuring single molecule transport across solid-state nanopores is similar to the one we described before.[8, 9] Briefly, a nanometer-sized hole is drilled into a 20 nm thin free-standing low-stress silicon nitride (SiN) membrane with a highly focused electron beam in a transmission electron microscope. The chip is then mounted in a microfluidic flow cell between two compartments filled with a monovalent KCl or NaCl or LiCl salt solution, with an additional 10 mM Tris-HCl and 1mM EDTA (TE) at pH 8.0 at room temperature. Below 100 mM the TE was diluted proportionally to the overall salt concentration. Subsequent application of an electric voltage across the membrane results in an ionic current through the pore, which is temporarily reduced upon the passage of a molecule. All experiments on bare DNA reported in this study were done on unmethylated double-stranded 48.5 kbp long λ-DNA (Promega, Madison, WI). RecA-λ-DNA filaments where λ-DNA is fully coated with RecA protein were formed as described before.[10] The event-fitting algorithm used to analyze the translocation events was as described before.[11] We used nanopores with diameters from approximately 5 to 25 nm in diameter, with linear I–V relations, and good low-frequency noise characteristics. Current traces at an applied voltage of 200 mV and below were measured at 10 kHz bandwidth using a resistive feedback amplifier (Axopatch 200B, Axon Instruments) and digitized at 100 kHz, or at 100 kHz bandwith and digitized at 500 kHz at voltages above 200 mV. All nanopores were treated in an oxygen plasma for about 30 s on both sides prior to use. All experiments were reproduced at least 3 times on similarly sized (15 – 23 nm in diameter) nanopores. The reported DNA translocation data were taken on a 21 nm pore; the RecA data were taken on a 23 nm pore.

3 RESULTS AND DISCUSSION

First, we present new measurements of DNA translocations at both high (1 M) and low (0.1 M) salt concentrations of KCl, NaCl, and LiCl for a broad range of voltages. Our results for KCl are consistent with earlier work on salt[12-14] and voltage[15, 16] dependence of dsDNA translocation. In previous work it was noticed that at low salt concentrations translocation of a DNA through a nanopore can lead to current enhancement.[13] Indeed, a transition from current decrease to current increase was discovered as salinity was increased in both nanochannels[14] and nanopores[12]. This ion current crossover can be understood from a simple model, where current decreases result from a reduced volume for ion flow due to the finite volume of the DNA strand, and current increases are attributed to an increased density of the counterions (cations) that screen the charge of the DNA backbone.[12] The voltage dependence of ΔG in KCl was reported before at both low[16] and high[15] salt concentrations. In 1 M KCl, it was found[15] that the current blockade value increased from 1.6 to 2.3 nS as the applied bias voltage increased from 100 to 600 mV. A similar effect was measured for dsRNA. The authors noted that these results suggest that the molecules were occupying a larger effective volume in or close to the nanopore at higher voltages. At 0.1 M KCl, it was reported[16] that at 200 mV DNA translocations produced upward pulses (current increases), whereas at 400 mV downward pulses were observed (current decreases). The authors attributed this to effect to a competition between the blocking current and the counterion current, where the latter saturated at high electric field.

The set of experiments described in this work confirms and expands the above mentioned observations. We further elucidate the role of counterions by doing measurements with different types of cations (Li$^+$, Na$^+$ and K$^+$; see also Ref. 9), and observe that the voltage V_0 where a crossover from current enhancements to current decreases occurs, decreases with cation mobility: V_0 (LiCl) $<$ V_0 (NaCl) $<$ V_0 (KCl). We discuss four possible scenarios to explain our data. Additionally, we present new data on the salt and voltage (non-)dependence of ΔG for RecA-DNA filaments, and discuss how it fits into the discussed scenarios.

We first discuss bare DNA translocations. Example traces of dsDNA translocations at several different voltages at low salt (0.1 M KCl) are shown in Figure 1. Upon addition of DNA molecules to the *cis* side, spikes appear in the traces. Each spike corresponds to the translocation of a single DNA molecule. Interestingly, at low voltages (300 mV and lower) the direction of the spikes is upward (i.e., the current increases), whereas at high voltages (400 mV and higher) we observe downward spikes (i.e., the current decreases).

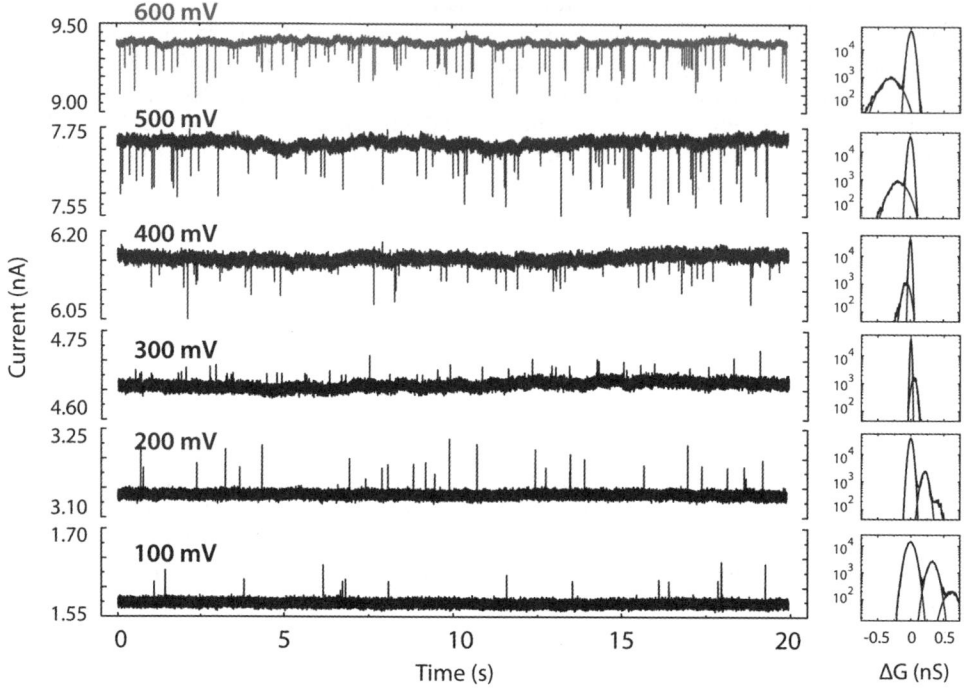

Figure 1 *Example current traces for DNA translocations at several different voltages (100 mV; 200 mV; 300 mV; 400 mV; 500 mV; 600 mV) for a nanopore of 21 nm in diameter at 0.1 M KCl. Spikes in the traces correspond to single-molecule translocation events. As is also clear from the conductance histograms shown on the right, at low voltages (below 350 mV) DNA translocation results in current enhancement, while at high voltages (above 350 mV) current blockades are seen.*

We analyse the data by making conductance histograms (on the right in Fig. 1) from which we extract the ΔG value at each particular voltage. We assign the center position of the fitted Gaussian as the value for ΔG. Results of the data analysis are shown in Figure 2a. Here, each point corresponds to a particular measurement condition (salt; molarity; voltage). We probed a wide range of salt concentrations (20 mM – 1 M) and voltages (10 – 1000 mV). A clear trend of increasing ΔG with voltage is found for DNA, both at 100 mM and at 1 M KCl, as well as for NaCl and LiCl. At high salt, we measure only current decreases, independent of voltage. At low salt concentrations, this leads to sign reversal of ΔG as a function of voltage for DNA translocations. Interestingly, we observe that the voltage V_0 at which the crossover occurs at low salt depends on the cation type (Fig. 2a), with $V_0(LiCl) = 110 \pm 7$ mV $< V_0(NaCl) = 202 \pm 3$ mV $< V_0(KCl) = 351 \pm 2$ mV.

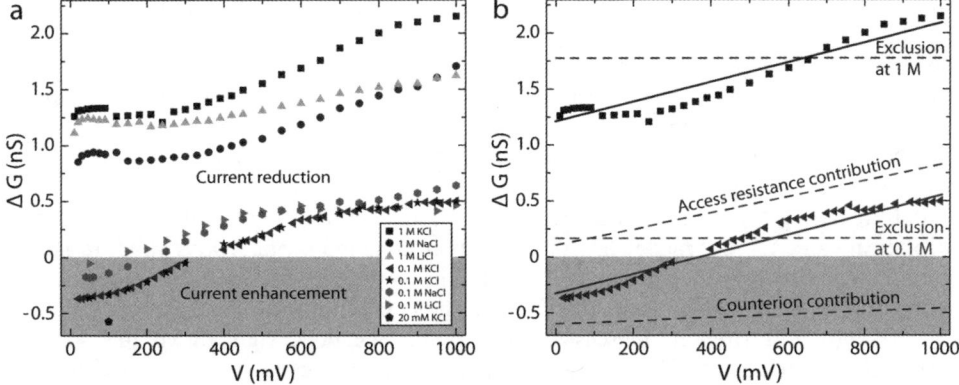

Figure 2 *(a) Voltage dependence of ΔG for dsDNA translocations in different solvents, as indicated in the legend. In all solvents a clear trend of increasing ΔG with voltage is found. At low salt, the crossover voltage is found to depend on the cation type. All data were taken on the same 21 nm nanopore, except the dataset at 0.1 M KCl indicated by blue stars, which was taken on an 18 nm nanopore. Positive ΔG signals a current decrease; negative ΔG a current enhancement. (b) Same data as in (a), but shown only for 0.1 M KCl (grey) and 1 M KCl (black). A simple model (scenarios (iii)+(iv) discussed in the text below) qualitatively predicts the observed behavior. Solid lines are the sum of the dotted lines, i.e., the volume exclusion, counterion, and access resistance contributions. The model assumes that both access resistance and counterion contributions increase linearly with voltage but are to first order independent of salt concentration, whereas the volume exclusion contribution is constant with voltage and scales linearly with salt concentration.*

Why does the value of ΔG increase with voltage? We discuss the following possible scenarios to explain this:

(i) Conformational changes within the DNA. For example, DNA might change from the B-duplex form to another form such as A-DNA. Or the bases might stick out at higher voltages and perhaps occupy a larger effective volume. These conformational changes seem however unlikely given the fact that quite a significant increase (of multiple nanometers) in effective DNA diameter would be required to explain the data, especially at low salt.

(ii) DNA stretching. We expect the force on a DNA molecule in a nanopore to exceed the threshold for overstretching the DNA (~60 pN) at around 400 mV for a 20 nm nanopore,[17] i.e. within the range of our experiments. Stretching the DNA will reduce its charge density compared to the case of unstretched DNA. This effect could lead to an increase in ΔG at higher voltages. However, the measured transition in ΔG from increase to blockade is very smooth, whereas – based on the force-extension curve in DNA stretching – one would instead expect a relatively sharp transition in this scenario.

(iii) Reduction of the counterion contribution. The change in conductance, ΔG, due to DNA translocation was expressed as follows[12] (where we adopted a -1 sign to be consistent with Figure 1, i.e., positive ΔG signals a current decrease; negative ΔG a current enhancement)

$$\Delta G = \frac{1}{L_{pore}} \frac{\pi}{4} d_{DNA}^2 \left(\mu_K + \mu_{Cl} \right) n_{KCl} e - \mu_K^* q_{l,DNA}^* \tag{1}$$

where d_{DNA} (2.2 nm) is the diameter of the molecule, L_{pore} is the length of the pore, n_{KCl} is the number density of potassium or chloride ions, e is the elementary charge, μ_K and μ_{Cl} are the electrophoretic mobilities of potassium and chloride ions, μ_K^* is the effective electrophoretic mobility of potassium ions moving along the DNA, and $q_{l,DNA}^*$ is the effective charge on the DNA per unit length, which, neglecting possible stretching, is assumed to be constant. Note that the first term in Eq. 1 leads to conductance decrease due to volume exclusion, while the second term can be associated with a conductance increase due to counterion flow.

It is tempting[12] to simply assume μ_K^* to be equal to its bulk value μ_K. However, theoretical work[18, 19] has made clear that this approximation is not always valid. In fact for strongly coupled polyelectrolyte (PE)-counterion systems the mobility of condensed counterions is negative for small fields, when the counterions are dragged along with the PE. As the field strength increases, the mobility changes sign above the so-called "saturation electric field" E_{sat}, the mobility is positive and the counterions glide along the PE. Experimental work (see Ref. 16 for an overview) on a 100 bp long DNA molecule found $E_{sat}=2\cdot10^4$ V/cm, which would correspond to 40 mV for our experimental case. This might explain the little bump between 0 – 150 mV in our high salt data in Figure 2a, where at low voltages (~0 – 50 mV) the counterions do not conduct, while at higher voltages (~50 – 150 mV) they start to glide along the DNA and produce an additional current. It is however unclear why this signature appears to be absent in the low salt data.

Finally, a reduction of 50%[19] up to nearly 100%[18] of the number of condensed counterions in the vicinity of the PE was observed as the electric field was increased, hence reducing $q_{l,DNA}^*$ at high fields. This effect will induce larger ΔG values for larger voltages. However, the magnitude of this effect is limited due to increased mobility of counterions μ_K^* at higher voltages (up to nearly their bulk values at high fields).[18, 19]

(iv) Access resistance. In this scenario the presence of the DNA in the access resistance region[8] becomes more pronounced at high voltages, as the DNA molecule is electrophoretically pulled closer towards the pore entrance. This is not captured in Eq. 1 where the access resistance is neglected. Recently, it was measured that the access resistance region is significant and extends up to about 100 nm away from the pore for a 10 nm pore.[20] Indeed, one cannot neglect the presence of a DNA molecule near the pore entrance, as this changes the access resistance. For example, an increase in ΔG with DNA contour length was measured,[21] where it was argued that the larger radius of gyration of the longer molecules increases the access resistance. Notably, this scenario would hold both at low and high salt, which fits with our observation that the effect of a change in ΔG vs V is of similar magnitude at both low and high salt (Figure 2a).

From the 4 scenarios discussed here, a combination of scenarios (iii) and (iv) seems most plausible. Figure 2b shows how a combination of these two scenarios can qualitatively explain the measured data, where for simplicity we assume that both access resistance and counterion contributions increase linearly with voltage but are independent of salt concentration, whereas the volume exclusion contribution is constant with voltage and scales linearly with salt concentration. Note that scenario (iii) in itself would be insufficient to explain the data (except if one would drop the, very plausible, assumption that volume exclusion should scale with molarity).

How can we understand the shift in crossover voltage for different counterions? The data presented in Figure 2a shows that the crossover voltage V_0 decreases with cation mobility, with $V_0(LiCl) = 110 \pm 7$ mV; $V_0(NaCl) = 202 \pm 3$ mV; $V_0(KCl) = 351 \pm 2$ mV. The relative bulk electrophoretic mobilities are $\mu_K : \mu_{Na} : \mu_{Li} = 1: 0.682 : 0.525$.[22] Returning to Eq. 1, we see that for both NaCl and LiCl the first term gives less volume exclusion blockade. The second term, however, predicts a lower contribution from counterions. At low salt, where the volume exclusion component is relatively small, the latter is the dominant contribution and hence we expect this to lead to an increase in ΔG at any fixed voltage, or in other words, to a shift of the curve to the left, i.e., a lower V_0.

Additionally, we present ΔG data for translocations of DNA that is fully covered with RecA (RecA-DNA filaments).[7, 10] Figure 3 shows (a) the voltage and (b) the salt dependence of ΔG for such filaments. We find that ΔG remains nearly constant in the range 0 – 800 mV. Interestingly, this behavior of ΔG is markedly different from that of dsDNA as well as ssDNA.[23] We speculate that this difference is caused by the stiffness of the molecules (see the discussion below and Figure 4). For the salt dependence we find that ΔG increases linearly with salt concentration. Extrapolating the (limited available) data at high salt to low salt concentrations, one would expect current increases below about 0.2 M KCl.

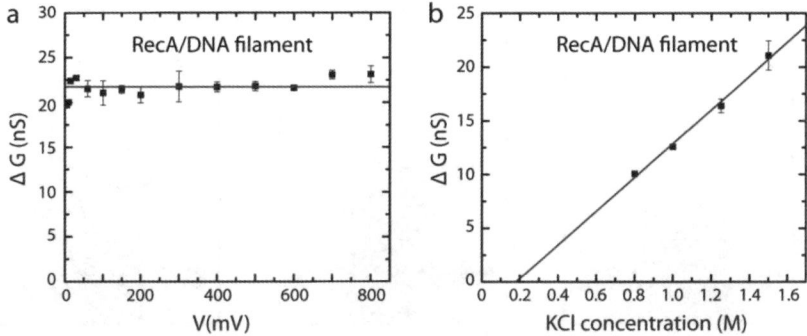

Figure 3 *(a) Voltage dependence of ΔG for RecA-DNA filaments translocations at 1.5M KCl. No strong dependence is found. Solid line is a constant fit to the data. (b) Salt dependence of ΔG for RecA-DNA filament translocation. Data were taken at 100 mV. Solid line is a linear fit to the data. A linear extrapolation of the data to low salt indicates a crossover from increases to decreases at ~ 0.2 M KCl, similar to the value found in static capture experiments.[24]*

Such a crossover was indeed obtained for RecA-DNA filaments in a nanopore optical tweezer experiment where the filaments were statically captured into the pore.[24] Here, a crossover from current decreases to increases was found to occur at 218 ± 81 mM KCl, i.e., close to our extrapolated value. It is interesting to note that for RecA-DNA filaments the absolute values of ΔG for translocation (11.4 ± 0.7 nS)[10] and static capture (7.5 ± 1.8 nS;[10] or, in a different study, 6.5 ± 0.9 nS)[24] experiments can differ by up to a factor 1.6. Following the discussion above (scenario (iv)), this discrepancy might be caused by the different conformations of the molecules at the *cis* side. Based on this argument one would expect a similar discrepancy for bare DNA molecules, which has not been reported. A fully satisfying explanation for the discrepancy between the RecA-DNA filament translocation and capture conductance blockade values remains lacking at the moment. It would be good to see a thorough theoretical treatment of the cation dynamics in both situations, which may shed light on this.

Let us now compare the access resistance scenario (iv) for three very different molecules: single-stranded DNA (ssDNA), double-stranded DNA (dsDNA) and RecA-DNA filaments. It is well know that ssDNA is very flexible (persistence length L_p of order 1 nm) and curls up into a small blob, comparable in size to the size of the access resistance region. dsDNA is stiffer ($L_p \approx 50$ nm), whereas RecA-DNA filaments are essentially straight rods ($L_p \approx 1$ μm)[25] on the scale of the nanopore. Interestingly, the voltage dependence of ΔG is particularly strong for (heterosequence) ssDNA ($\Delta G \approx aV^b$ with $a = 2.0 \pm 0.3$ nS and $b = 0.37 \pm 0.03$ and V the voltage in mV),[23] where the increase in ΔG can be associated with a more prominent presence of the ssDNA in the access resistance region. As depicted schematically in Figure 4b, this effect is less strong – but still present – for dsDNA. For the stiff RecA-DNA filament the majority of the filament is far away from the pore, hence the influence of an increased voltage is mainly felt by the relatively small part of the filament near the pore, leading to a nearly constant ΔG over a broad range of voltages (0 – 800 mV; see Figure 3a). This provides additional evidence for scenario (iv).

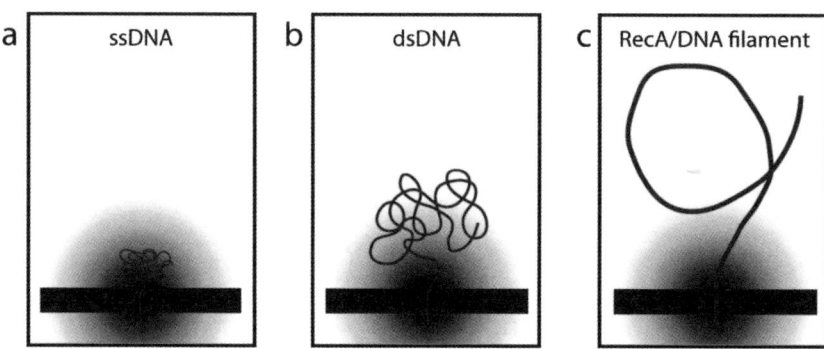

Figure 4 *Schematic for (a) ssDNA, (b) dsDNA, and (c) RecA-DNA filament translocation. Gray semi-spheres indicate the approximate extension of the access resistance region. The ssDNA is fully inside the access resistance region, while the dsDNA is partly in this region and the RecA-DNA filament is mainly far away from this region.*

Finally, we would like to note that for small pores (<10 nm) and at intermediate voltages, we observed a new type of intriguing composite events for dsDNA translocations, where the current first decreases and then increases within each single translocation event.[26]

4 CONCLUSION

We have presented new measurements of the conductance blockade (ΔG) induced by DNA and RecA-DNA filament translocation in a wide range of salt concentrations and voltages. In all measurement conditions, a clear trend of increasing ΔG with voltage was found for DNA, while ΔG was found to remain independent of voltage for RecA-DNA filament transloations. At low salt concentrations, a sign reversal of ΔG as a function of voltage was observed for DNA translocations. We find that the most probable scenario to explain all data is one where the contribution due to the access resistance increases with voltage due to the presence of a molecule near the pore entrance. However, the effects discussed in the other scenarios, such as reduced counterion flow, are not mutually exclusive with this scenario and can play a role as well.

5 ACKNOWLEDGEMENTS

We thank A. Aksimentiev, T.R. Blosser, N.H. Dekker, A.R. Hall, M. van den Hout, X.J.A. Janssen, P.M. Jonsson, S.G. Lemay, C. Plesa, G.F. Schneider, and G.V. Soni for discussions. This work is supported by the EC program NanoSci-E+, the European Union's Seventh Framework Program (FP7/2007-2013) under grant agreement no. 201418 (READNA), and ERC-2009-AdG Grant 247072 NANOforBIO.

References

1 C. Dekker, *Nat. Nanotechnol.*, 2007, **2**, 209.
2 S. Howorka and Z. Siwy, *Chem. Soc. Rev.*, 2009, **38**, 2360.
3 A. Aksimentiev, *Nanoscale*, 2010, **2**, 468.
4 U.F. Keyser, *J. R. Soc. Interface*, 2011, **8**, 1369.
5 B.M. Venkatesan and R. Bashir, *Nat. Nanotechnol.*, 2011, **6**, 615.
6 S.W. Kowalczyk, T.R. Blosser and C. Dekker, *Trends Biotechnol.*, 2011, **29**, 607.
7 S.W. Kowalczyk, A.R. Hall and C. Dekker, *Nano Lett.*, 2010, **10**, 324.
8 S.W. Kowalczyk, A.Y. Grosberg, Y. Rabin and C. Dekker, *Nanotechnology*, 2011, **22**, 315101.
9 S.W. Kowalczyk, D.B. Wells, A. Aksimentiev and C. Dekker, *Nano Lett.*, Article ASAP, 2012, DOI: 10.1021/nl204273h.
10 R.M.M. Smeets, S.W. Kowalczyk, A.R. Hall, N.H. Dekker and C. Dekker, *Nano Lett.*, 2009, **9**, 3089.
11 A.J. Storm, C. Storm, J. Chen, H. Zandbergen, J. Joanny and C. Dekker, *Nano Lett.*, 2005, **5**, 1193.
12 R.M. Smeets, U.F. Keyser, D. Krapf, M.Y. Wu, N.H. Dekker and C. Dekker, *Nano Lett.*, 2006, **6**, 89.
13 H. Chang, F. Kosari, G. Andreadakis, M.A. Alam, G. Vasmatzis and R. Bashir, *Nano Lett.*, 2004, **4**, 1551.
14 R. Fan, R. Karnik, M. Yue, D. Li, A. Majumdar and P. Yang, *Nano Lett.*, 2005, **5**, 1633.
15 G.M. Skinner, M. van den Hout, O. Broekmans, C. Dekker and N.H. Dekker, *Nano Lett.*, 2009, **9**, 2953.
16 H. Chang, B.M. Venkatesan, S.M. Iqbal, G. Andreadakis, F. Kosari, G. Vasmatzis, D. Peroulis and R. Bashir, *Biomed. Microdevices*, 2006, **8**, 263.
17 S. van Dorp, U.F. Keyser, N.H. Dekker, C. Dekker and S.G. Lemay, *Nature Physics*, 2009, **5**, 347.
18 R.R. Netz, *J. Phys. Chem. B*, 2003, **107**, 8208.
19 K. Grass and C. Holm, *Soft Matter,* 2009, **5**, 2079.
20 C. Hyun, R. Rollings and J. Li, *Small*, 2011, DOI: 10.1002/smll.201101337.
21 M. Wanunu, J. Sutin, B. McNally, A. Chow and A. Meller, *Biophys. J.*, 2008, **95**, 4716.
22 D.R. Lide, *CRC Handbook of Chemistry and Physics*, 85th ed.; internet version, 2005.
23 S.W. Kowalczyk, M.W. Tuijtel, S.P. Donkers and C. Dekker, *Nano Lett.*, 2010, **10**, 1414.
24 A.R. Hall, S. van Dorp, S.G. Lemay and Cees Dekker, *Nano Lett.*, 2009, **9**, 4441.
25 M. Hegner, S.B. Smith and C. Bustamante, *Proc. Natl. Acad. Sci. U.S.A.,* 1999, **96**, 10109.
26 S.W. Kowalczyk and C. Dekker, in preparation.

BIOCHEMICAL SENSING WITH CHEMICALLY MODIFIED SYNTHETIC ION CHANNELS

M. Ali,*[,1,2] S. Nasir,[1,2] Q.H. Nguyen,[1,2] R. Neumann[2] and W. Ensinger[1,2]

[1] Technische Universität Darmstadt, Fachbereich Material-u. Geowissenschaften, Fachgebiet Materialanalytik, D-64287 Darmstadt, Germany.
[2] GSI Helmholtzzentrum für Schwerionenforschung, D-64291 Darmstadt, Germany.

1 INTRODUCTION

Inspired from the working and functionality of natural ion channels in living organisms, scientists and engineers have tried to miniaturize electrochemical devices exhibiting potential applications in various biochemical sensing[1-6] and separation processes.[7-9] Ion channels in a cell membrane regulate the flow of ions and molecules across the cell boundaries.[10, 11] Majority of these ion channels are selective for a given ion or molecule, and are responsive so that the transport is regulated by changes in the electrical potential difference/pressure, or by the binding of a chemical to the channel structure. Natural ion channels have also become a model system to investigate transport phenomena at the nanoscale level.[10, 11]

Synthetic nanochannels fabricated in solid-state materials present several advantages over their biological counterparts such as robustness, control over shape and size, and tailorable surface charge properties.[12, 13] Furthermore, synthetic ion channels have attracted a considerable interest due to their unique transport properties and potential applications.[14-17]

During the recent years, track-etched single conical nanochannels have been proven a novel biosensing platform on the basis of their ionic transport properties, *i.e.*, ion current rectification and permselectivity, which depends sensibly on surface charge density and effective diameter, especially at the tip opening of the channel.[14-17] Any minor change in the surface charge and/or effective cross-section can lead to a significant change in electronic read-out originated from the ionic transport through the channel. To date, several approaches have been pursued to achieve active control over ion transport in nanoconfined geometries by decorating the channel surface with a variety of chemical functionalities that respond to different external stimuli.[18-22]

Here we would like to present two types of nanoporous sensors based on single conical polymer nanochannels for the sensitive detection of hydrogen peroxide (H_2O_2) and single-stranded DNA nucleotides. For the case of H_2O_2 sensor, the surface and inner channel walls are decorated with the horseradish peroxidase (HRP) enzyme. The immobilized enzyme remains active in redox reactions that occur inside a single nanochannel in the presence of even nano-molar concentrations of H_2O_2. Monitoring the redox reactions *via* changes of the nanochannel transport properties offers an easy method to detect H_2O_2. For the second type sensor, channel walls are functionalised with a peptide nucleic acid (PNA), which acted as binding sites for the capturing of complementary DNA oligomer. The hybridisation process between PNA and DNA boosted the channel surface charge density due to the negatively charged phosphate backbone of DNA oligomer. Upon DNA/DNA duplex formation, the sensor directly transduces the hybridization events into a measurable electronic read-out signal.

2 METHOD AND RESULTS

2.1. Fabrication of Single Conical Nanochannels

Single conical nanochannels were fabricated in ion tracked polymer membranes by the asymmetric track-etching technique developed by Apel and co-workers.[23] The conical shape of the channels resulted from the asymmetric development of the damaged tracks in an etching solution. The ion tracks in polyethylene terephthalate (PET) and polyimide (PI) membranes are chemically etched with sodium hydroxide (9M NaOH) at room temperature and sodium hypochlorite (NaOCl) at 50°C, respectively.[24]

2.2. Functionalisation of Channel Surface

As a result of the heavy ion irradiation and chemical etching process, carboxyl (COOH) groups are generated on the membrane surface. These groups were first activated in a water solution of N-(3-dimethylaminopropyl)-N'-ethylcarbodiimide (EDC) and N-hydroxysulfosuccinimide (sulfo-NHS).[25] The activating reagent EDC first converts the COOH groups into a highly reactive o-acylisourea intermediate. This intermediate was further converted into a more stable sulfo-succinimidyl ester in the presence of excess sulfo-NHS molecules. Then, the succinimidyl intermediate is covalently coupled with the primary amine groups in HRP molecules[3] and terminal amine groups of PNA (H$_2$N–O–Lys–TAGTCGGAAGCA) probe to form stable amide bonds (Scheme 1).[1]

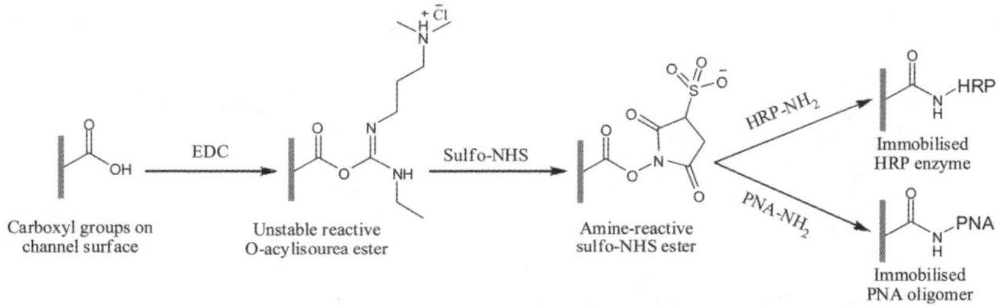

Scheme 1. *Reaction scheme for the covalent immobilisation of HRP-enzyme molecules and PNA-oligomers on the inner channel walls via carbodiimide coupling chemistry.*

2.3. HRP-modified nanochannels and hydrogen peroxide sensing

Figure 1 shows the pH-dependant I–V characteristics of a single conical nanochannel prior to and after the enzyme immobilization. It is known that HRP has a molecular weight of approximately 44 kDa and ~ 3.6 nm in diameter[26], thus its attachment into our ~ 20 nm in diameter channel was expected to diminish the effective channel opening, resulting in the lowering of ionic current as shown in Fig.1b. Because of the amphoteric nature of immobilised HRP enzyme, the ionic transport properties of the modified channel strongly depend on the pH value of the electrolyte solution. At pH = 6.5, the HRP-modified channel still exhibits ionic current rectification characteristics. On the contrary, in acidic solution (pH = 3.0), the I–V curve is flipped so that the values of negative currents became higher than of positive currents (Fig. 1b). This inversion of rectification for the modified channel clearly showed the switching of the surface charge from negative (pH 6.5) to positive (pH 3.0) due to the protonation of amine groups on the HRP-molecules.[3]

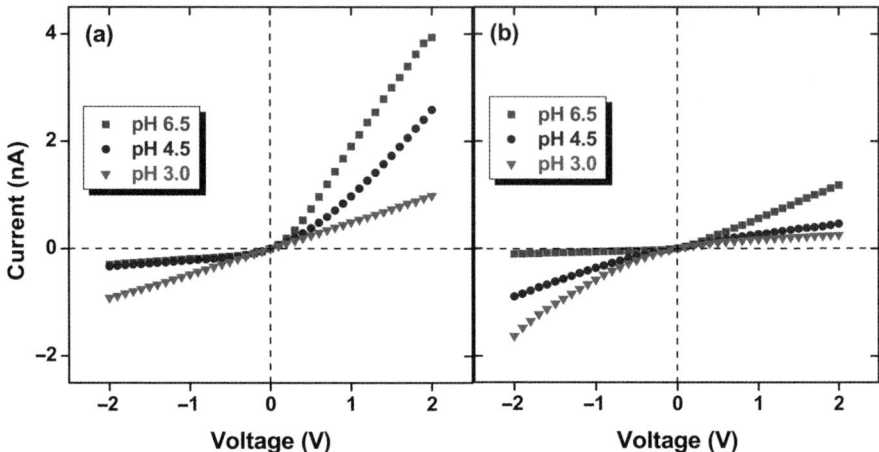

Figure 1. *Current-voltage characteristics of a single conical nanochannel in 0.1 M KCl buffered to different pH values as indicated in the Figure (a) before and (b) after*

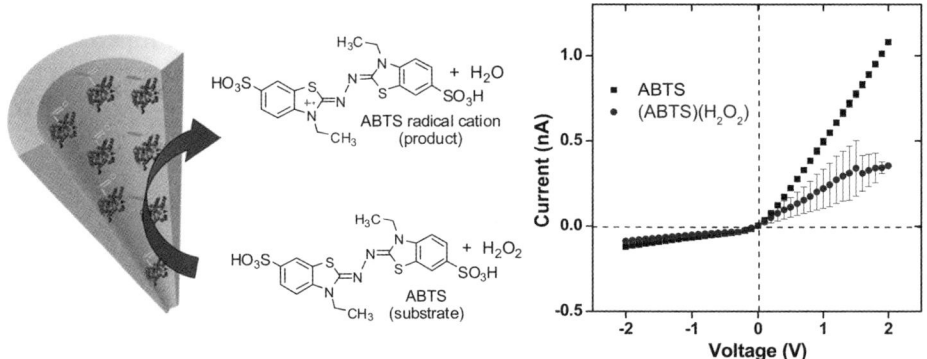

Figure 2. *Scheme of the bio-catalyzed redox reaction (left) of the immobilized HRP enzyme leading to the conversion of ABTS into respective radical cation product in the presence of hydrogen peroxide. I–V characteristics of a single HRP-modified channel (right) recorded in the absence and in the presence of 0.5 mM hydrogen peroxide in 0.1 M KCl solution containing 1.5 mM ABTS substrate.*

The next step is to investigate the activity of the immobilized enzyme in redox reactions, carried out in the confined volume of the channel. The HRP-nanochannel system is tested in a system containing 2,2′-azino-bis(3-ethylbenzothiazoline-6-sulfonate) (ABTS) as a substrate along with H_2O_2 as an analyte. Figure 2 (right) shows the transmembrane current through a single HRP-modified channel, recorded in the presence of the substrate ABTS (1.5 mM) dissolved in 0.1 M KCl (pH = 6.5) prior to and after the addition of H_2O_2. In the absence of H_2O_2, I–V curves were smooth and very similar to those recorded in just KCl solution. Addition of 0.5 mM H_2O_2 induced a significant decrease of positive currents, which also became more unstable. For positive voltages, the measured *I–V* curves additionally exhibited a hysteresis: the currents for voltages increasing from 0 to +2 V were significantly higher than the recordings for the voltage sweep from +2 to 0 V (Fig.2). We attributed the current changes to the appearance of cationic (ABTS radical cation) products of the redox reaction that occurred in the presence of HRP, ABTS and H_2O_2.[27, 28] We have assumed that the reaction products (ABTS$^{•+}$) obstructed the ion current flowing through the channel.[3]

2.4. PNA-modified nanochannel and recognition of DNA oligomer

For the case of PNA-modified channel, Figure 3 shows *I–V* curves of a single conical nanochannel having carboxyl, immobilised PNA probe and PNA/DNA duplex on the inner walls. Umodified channel exhibited cation selectivity with rectified ion current of 53.6 nA flowing across the single-channel membrane (Figure 3a). After immobilisation of PNA, a significant decrease in the rectified ionic current (form 53.6 to 16.7 nA) was observed under the same applied bias (Figure 3b). The change in the rectified ion current prior to and after modification is attributed due to the uncharged back-bone of immobilised PNA probe that confirmed the success of modification reaction.[1]

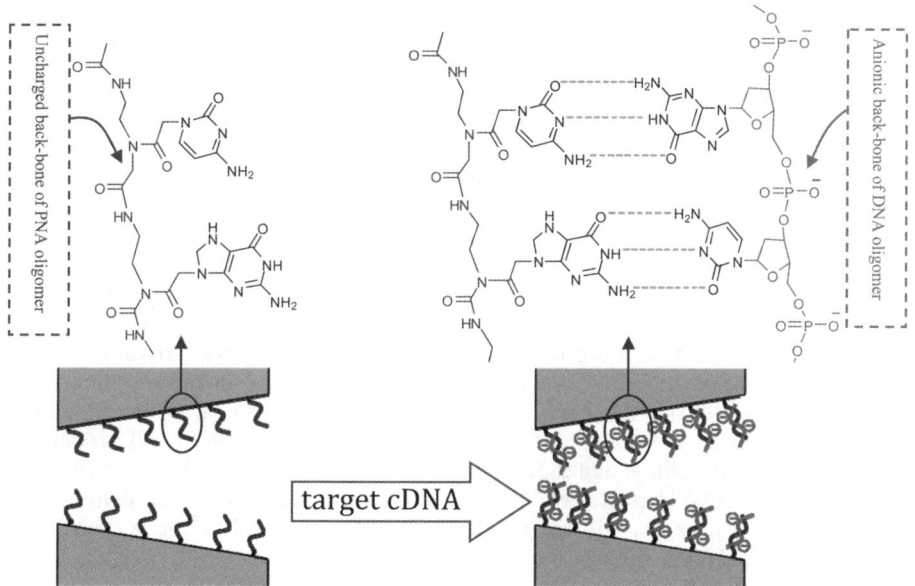

Scheme 2. Scheme describing the immobilised PNA-probe, and subsequent hybridization of complementary DNA oligonucleotide with the tethered PNA probe inside the single conical nanopore.

The PNA-modified channel as a biosensor was investigated to detect DNA oligomer 5'-TGCTTCCGACTA-3' (c-DNA) with sequence complementary to the immobilized PNA probe. It is known that PNA oligomer is an analogue of DNA, used as a DNA hybridization partner due to its chemical robustness. The hybridization of c-DNA strand on the sensor surface can be recognized from the changes in the rectified ion flux through the channel. It is worth mentioning that upon PNA/DNA hybridization, negative charges were generated on the inner channel wall due to the negatively charged phosphate backbone of the DNA nucleotide. Therefore, the presence of these negative charges triggered the ion current rectification, and the value of ionic current measured at +1 V was increased from 16.7 to 37.0 ± 3.2 nA as shown in figure 3c. From the *I–V* curve, indicating pronounced rectification, we conclude that the DNA recognition *via* PNA/DNA hybridization rendered the surface negatively charged, thus restoring the rectification characteristics displayed by the channel with the preferential transport of cationic species.

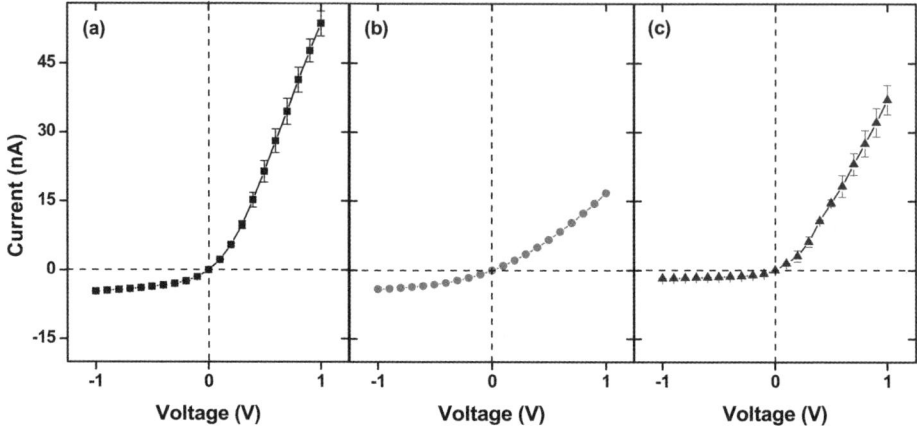

Figure 3. *I–V* characteristics of a single conical nanochannel bearing carboxylate groups (a), immobilized PNA probe (b), and PNA/DNA duplexes (c) on the channel surface, respectively.

Rectification properties of the nanochannels can be quantified by calculating the rectification degree (f_{rec}) defined as the ratio of absolute values of ion currents measured at voltages of the same amplitude but opposite polarities. The value of f_{rec} is sensibly correlated with the surface charge, and consequently a slight increase/decrease in surface charge can trigger a marked increase/decrease in rectification ratio of the nanochannel. It is evident from *I–V* curves shown in Figure 3 that immobilization of PNA can lead to ~4-fold decrease in rectification ratio from 11 to 4, measured at a potential of ±1 V. After hybridizing DNA onto the PNA-modified channel, an increase in the rectification ratio from ~4 to 21 (~8-fold) is observed. Compared with the as-prepared nanochannel, the rectification ratio is almost doubled in the case of a channel bearing PNA/DNA duplexes. This clearly shows that the magnitude of surface charges is increased upon hybridization because of the negative character of DNA molecules. Similarly, Fu *et al.* demonstrated a nanopore DNA sensor based on modified nanopipettes. They have showed that electrostatically attached single-stranded DNA oligomer selectively hybridizes with complementary DNA oligomer, resulted an increase in the rectification ratio displayed by DNA-modified nanopipettes pores.[29] This further supports our findings that introducing negatively charged DNA strands, *via* hybridization onto the PNA surface, led to a significant increase in the rectification ratio.

3 CONCLUSIONS

In conclusion, we have demonstrated the decoration of inner channels walls with biomolecules (HRP enzyme and PNA oligomers) *via* exploiting the inherent functional groups on the channel surface. The HRP- and PNA-functionalized single nanochannels were successfully employed for the biochemical sensing of H_2O_2 and cDNA oligomers, respectively. For the case of HRP-channel system, the cationic products (ABTS$^{\cdot+}$) of the redox reaction occurring inside the confined geometry resulted in the reduction of ionic current in a voltage-dependent fashion. For the case of PNA-modified channel, surface charge-dependent *I–V* characteristics exhibited the PNA/DNA hybridization process inside the constrained geometry of the nanofluidic device. For both sensors, the detection signal relies on the *I–V* characteristics which are much less susceptible to external noise compared to that of the time-resolved sensing of biochemical analyte molecules.

ACKNOWLEDGMENTS

The authors gratefully acknowledge financial support by the Beilstein-Institut, Frankfurt/Main, Germany, within the research collaboration NanoBiC, and thanks Prof. C. Trautmann for support with irradiation experiments.

References

1. M. Ali, R. Neumann and W. Ensinger, *ACS Nano*, 2010, **4**, 7267-7274.
2. M. Ali, B. Schiedt, R. Neumann and W. Ensinger, *Macromol. Biosci.*, 2010, **10**, 28-32.
3. M. Ali, M. N. Tahir, Z. Siwy, R. Neumann, W. Tremel and W. Ensinger, *Anal. Chem.*, 2011, **83**, 1673-1680.
4. C. P. Han, X. Hou, H. C. Zhang, W. Guo, H. B. Li and L. Jiang, *J. Am. Chem. Soc.*, 2011, **133**, 7644-7647.
5. S. Howorka and Z. Siwy, *Chem. Soc. Rev.*, 2009, **38**, 2360-2384.
6. Z. Siwy, L. Trofin, P. Kohli, L. A. Baker, C. Trautmann and C. R. Martin, *J. Am. Chem. Soc.*, 2005, **127**, 5000-5001.
7. M. Ali, S. Nasir, P. Ramirez, I. Ahmed, Q. H. Nguyen, L. Fruk, S. Mafe and W. Ensinger, *Adv. Funct. Mater.*, 2012, **22**, 390-396.
8. M. Nishizawa, V. P. Menon and C. R. Martin, *Science*, 1995, **268**, 700-702.
9. E. N. Savariar, K. Krishnamoorthy and S. Thayumanavan, *Nat. Nanotechnol.*, 2008, **3**, 112-117.
10. E. Gouaux and R. MacKinnon, *Science*, 2005, **310**, 1461-1465.
11. B. Hille, *Ionic channels of excitable membranes*, Sinauer Associates Inc., Sunderland, MA, 2001.
12. C. Dekker, *Nat. Nanotechnol.*, 2007, **2**, 209-215.
13. Z. S. Siwy and S. Howorka, *Chem. Soc. Rev.*, 2010, **39**, 1115-1132.
14. P. Ramirez, P. Y. Apel, J. Cervera and S. Mafe, *Nanotechnology*, 2008, **19**, 315707.
15. Z. S. Siwy, *Adv. Funct. Mater.*, 2006, **16**, 735-746.
16. D. Stein, M. Kruithof and C. Dekker, *Phys. Rev. Lett.*, 2004, **93**, 035901.
17. I. Vlassiouk and Z. S. Siwy, *Nano Lett.*, 2007, **7**, 552-556.
18. M. Ali, S. Mafe, P. Ramirez, R. Neumann and W. Ensinger, *Langmuir*, 2009, **25**, 11993-11997.
19. M. Ali, P. Ramirez, S. Mafe, R. Neumann and W. Ensinger, *ACS Nano*, 2009, **3**, 603-608.
20. M. Ali, B. Schiedt, K. Healy, R. Neumann and W. Ensinger, *Nanotechnology*, 2008, **19**, 085713.
21. M. Ali, B. Yameen, R. Neumann, W. Ensinger, W. Knoll and O. Azzaroni, *J. Am. Chem. Soc.*, 2008, **130**, 16351-16357.
22. X. Hou, W. Guo and L. Jiang, *Chem. Soc. Rev.*, 2011, **40**, 2385-2401.
23. P. Y. Apel, Y. E. Korchev, Z. Siwy, R. Spohr and M. Yoshida, *Nucl. Instrum. Methods Phys. Res., Sect. B*, 2001, **184**, 337-346.
24. Z. Siwy, Y. Gu, H. A. Spohr, D. Baur, A. Wolf-Reber, R. Spohr, P. Apel and Y. E. Korchev, *Europhys. Lett.*, 2002, **60**, 349-355.
25. G. T. Hermanson, *Bioconjugate Techniques*, Academic Press, San Diego, 1996.
26. A. E. Radi, X. Munoz-Berbel, M. Cortina-Puig and J. L. Marty, *Electroanalysis*, 2009, **21**, 1624-1629.
27. E. N. Kadnikova and N. M. Kostic, *J. Mol. Catal. B: Enzym.*, 2002, **18**, 39-48.
28. J. N. Rodriguez-Lopez, D. J. Lowe, J. Hernandez-Ruiz, A. N. P. Hiner, F. Garcia-Canovas and R. N. F. Thorneley, *J. Am. Chem. Soc.*, 2001, **123**, 11838-11847.
29. Y. Q. Fu, H. Tokuhisa and L. A. Baker, *Chem. Commun.*, 2009, 4877-4879.

LABEL-FREE SCREENING OF NICHE-TO-NICHE VARIATION IN SATELLITE STEM CELLS USING FUNCTIONALIZED PORES

M. R. Chapman [1], K. Balakrishnan[2], M. J. Conboy[3], S. K. Mohanty[2], E. Jabart[3], J. Li[3], H. Huang[4], J. Hack[2], I. Conboy[3], L. L. Sohn[2]

[1]Biophysics Graduate Group, University of California, Berkeley, CA, 94720, USA
[2] Department of Mechanical Engineering, University of California, Berkeley, CA, 94720, USA
[3]Department of Bioengineering, University of California, Berkeley, CA, 94720, USA
[4]Department of Statistics, University of California, Berkeley, CA, 94720, USA

1 INTRODUCTION

Understanding embryonic organogenesis and adult tissue regeneration relies on the ability to isolate and characterizing stem cells. However, these cells are difficult to study because they constitute minute populations in organ niches and express multiple cell-surface markers, few of which are identified. Furthermore, the properties of these cells change quickly in vitro and possibly even during isolation procedures[1,2,3]. Perturbations introduced during sample isolation and processing, and the difficulty in determining gene-expression levels accurately in the low starting stem-cell numbers within the micro-anatomical niche further add further complications. Because they are better suited for large numbers of cells, fluorescence-activated cell sorting (FACS) and magnetic-activated cell sorting (MACS) do not easily enable niche-specific characterization or even separate hind-leg muscle groups in the case of muscle (satellite) stem cells. Microscopy, although capable of imaging stem cells in their niches, is labor intensive and quantifying expression levels is difficult. Adding to the overall complexity is the fact that FACS, MACS, and fluorescence microscopy depend on irreversible antibody binding to stem-cell surface proteins, potentially altering cell properties, including gene expression and regenerative capacity[4]. To address these challenges, we have developed a unique, label-free method for the objective, quantitative screening and characterization of single, functional organ stem cells. We demonstrate the power of our method by quantitatively screening satellite cells directly isolated from single muscle fibers.

2 METHOD AND RESULTS

For our measurements, single satellite cells, isolated from individual myofibers of extensor digitorum longus (EDL) muscle were injected into a polydimethylsiloxane (PDMS) pore (Figure 1a) that has been functionalized with a saturating concentration of either a specific or an isotype-control antibody. A non-pulsatile pressure[5] was used to drive single satellite cells through filters, an inner reservoir, and finally through the functionalized pore for measurement. As individual satellite cells transit the microchannel, the flow of current through the microchannel is partially blocked, leading to a transient increase, or pulse, in the electrical resistance (Figure 1b) that is subsequently measured and analyzed to characterize the cell[5,6,7,8,9,10,11]. The pulse magnitude and width correspond to cell size and transit time, τ, across the microchannel, respectively. Because of specific interactions, cells transiting a microchannel functionalized with an antibody specific for an expressed surface marker have longer transit times than those transiting an isotype-control pore or an unfunctionalized pore.

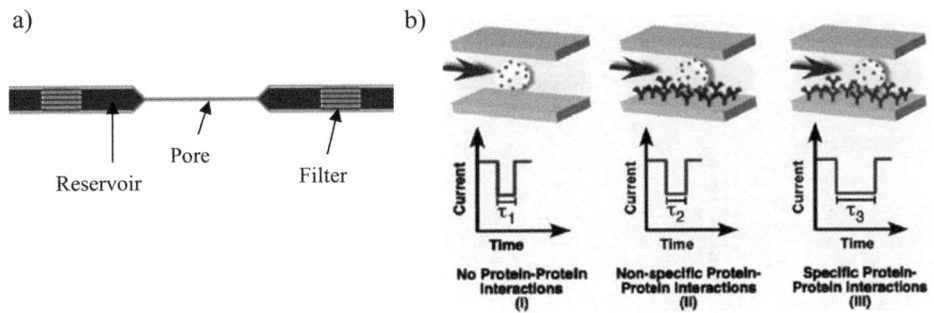

Figure 1 *Detailed view and description of the stem-cell analysis platform. a) A microchannel with two reservoirs, filters and a pore (800 μm x 25 μm x 25 μm, L x W x H). b) Pulse magnitude and width correspond to cell size and transit time, τ, respectively. When the pore is functionalized with antibodies that have a high affinity to a particular cell-surface epitope, transient binding between the two leads to a longer transit time (III) than that due to non-specific interactions in an isotype-control antibody pore (II) or an unfunctionalized pore (I)* (Carbonaro et al., 2008)[5].

To show that our method is capable of quantitatively determining both the expression and lack of expression of a particular surface antigen in a population, we screened primary-culture mouse myoblasts for Sca-1 and M-cadherin in prepared microchannels. In agreement with previous reports that were based on FACS analysis, we found that 2.7% of the cultured myoblasts were Sca-1+[12] and 93.0% were M-cadherin+[13,14,15] (Figure 2). To demonstrate that our method could screen a heterogeneous population, we screened cultured Sca-1+ cells and determined that 67.6% were Sca-1+. This was in excellent agreement with FACS, which had determined the Sca-1+ population to be 66.1%.

To determine if the transient binding between the functionalized antibody and cell-surface receptors activates receptor signaling, and therefore alters cell properties (a concern for FACS and MACS[4]) we analyzed whether Notch would become activated when freshly harvested satellite cells transited these pores. The anti-Notch1 antibody employed in our device has been shown to mimic the native ligand binding and activate Notch1 robustly in satellite cells, resulting in high levels of the truncated intracellular portion of Notch that is localized to the cell nucleus[16]. As a control, satellite cells also transited pores functionalized with IgG$_1$ control antibody. As additional controls, unscreened satellite cells were plated on IgG$_1$ and anti-Notch1 antibody-coated culture wells overnight in mitogen-low medium. To determine whether the transient interactions between the extracellular portion of the receptor and the functionalized antibodies activated the Notch pathway, we performed immunofluorescence on all cells using an antibody that specifically recognizes the truncated-activated form of Notch1. In contrast to wells coated with anti-Notch1 antibody, which induced robust nuclear-active Notch, cells from the IgG$_1$ and anti-Notch1 antibody pores and the IgG-coated wells showed low levels of Notch activation. Thus, the transient binding between functionalized antibody and specific receptors in our pore does not significantly contribute to Notch-pathway activation in satellite cells and provides strong evidence that changes in signal transduction and cell behavior should not be expected when cells are screened with our method. Therefore, in addition to screening rare cells, our label-free screening method is of general utility for the analysis of any cell population without altering cell properties or interfering with downstream assays or experiments.

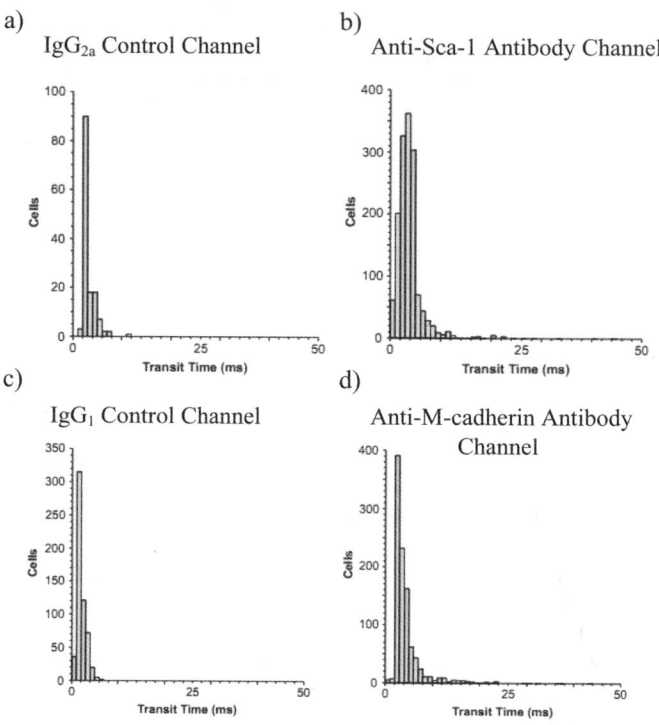

Figure 2 *Screening primary-culture mouse myoblasts with functionalized pores. a) Histogram of transit-times for myoblasts screened with IgG_{2a} isotype-control (n = 141, τ = 3.16 ± 0.13 ms) and b) anti-Sca-1 (n = 1469, τ = 3.89 ± 0.35 ms) antibody-coated microchannels, respectively. Using a False Discovery Rate (FDR) analysis, 2.7% of cells screened were found to be Sca+ c) Histogram of transit-times for myoblasts screened with IgG_1 isotype-control (n = 571, τ = 2.00 ± 1.01 ms) and d) anti-M-cadherin (n = 1004, τ = 4.45 ± 3.83 ms) antibody coated pores, respectively. Using a False Discovery Rate (FDR) analysis, 93.0% were found to be M-cadherin+ (13.2% high, 30.8% medium, and 49.0% low expression).*

Unlike bulk screening methods, which do not discriminate between cells residing in different individual myofibers or even elsewhere in the tissue, our method uniquely allows for the analysis of rare cells from a single myofiber niche. Fiber-to-fiber heterogeneity was examined for satellite-cell markers: Sca-1, CXCR4, β1-integrin, and M-cadherin. Muscle satellite cells were freshly isolated from single muscle fibers of EDL muscle from a 6 month old C57/black-6 mouse and injected into our devices for screening.

To determine which cells were Sca-1+, M-cadherin+, β1-integrin+, or CXCR4+, we performed a Dixon's Q Test on each of the cells transiting the specific antibody pores to assess whether a particular cell had an outlying slow transit time as compared to those cells passing through the isotype control pore. This is a robust statistical test used to identify values that diverge from a control sample and is ideal for small sample sizes[17,18,19]. Because there is a direct correlation between the density of available epitopes and transit time, we can are able to determine expression levels of Sca-1, M-cadherin, β1-integrin, and CXCR4. We thus used a Dixon's Q Test with conservative Bonferroni corrected significance levels of p-value cutoffs to determine the levels of expression of Sca-1, M-cadherin, β1-integrin, or CXCR4. As shown in the summary of experiments for each marker (Figure 3), some fibers have almost no (<5%) Sca-1, CXCR4, or β1-integrin expressing satellite cells, while others have a significant number (>50%). In contrast to Sca-1, CXCR4, and β1-integrin, M-cadherin showed the least variability, with ~35% of cells showing some level of expression in all three samples screened. The relatively consistent expression levels of M-cadherin that we measured suggest that the heterogeneity for the other markers is not the result of a variation in measurement; rather, it is an accurate reflection of the expression of these markers, themselves.

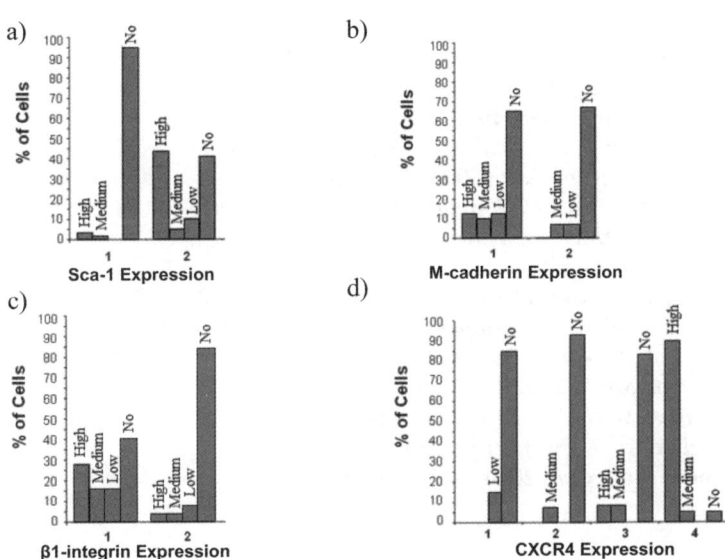

Figure 3 *a-d) Summary of the expression levels of Sca-1, M-Cadherin, β1-integrin, CXCR4 for different experiments corresponding to different muscle fibers. Wide heterogeneity in expression levels is evident for each surface marker.*

3 CONCLUSION

Using our novel microfluidics approach, we have shown a remarkable heterogeneity in all cell surface markers studied (Sca-1, CXCR4, M-cadherin, and β1-integrin) in satellite cells isolated from different single myofibers, which emphasizes the notion of complexity of the muscle stem-cell population. We also measured the surface expression of Notch1 on satellite cells without activating the Notch pathway using our innovative label-free screening technique. While satellite-cell heterogeneity has been previously suggested[5], this work reveals significant differences in the composition of satellite cells between different single myofibers derived from EDL; moreover, the data define satellite cell markers in an objective and quantitative way. Such results cannot be obtained with bulk sorting of satellite cells, which does not discriminate between cells purified from separate myofibers or even from separate hind leg muscle groups. Our screening and analysis method is also much more accurate as compared to manual microscopic evaluation of immunostained single fibers, yielding objective quantitative data (time-of-flight) for single live cells transiting across a surface functionalized with a saturating concentration of immobilized antibodies. As such, the method presented here combines the best of both techniques: the high-resolution of flow cytometry with the ability to analyze quantitatively rare stem cells that are associated with single muscle fibers

References

1   S. Kuci, et al., *Current Stem Cell Res. Ther.*, 2009, **10**, 107-117.
2   G. Shefer and Z. Yablonka-Reuveni, *Methods Mol. Biol.*, 2005, **290**, 281-304.
3   J. Slack, *Science*, 2008, **322**, 1498-1501.
4   A. Tarnok, H. Ulrich, and J. Bocsi, *Cytometry A*, 2010, **77**, 6-10.
5   A. Carbonaro, et al., *Lab Chip*, 2008, **8**, 1478-1485.
6   W. Coulter, *US Patent*, 1953, 2,656,508.
7   R. Deblois and C. Bean, *Rev. Sci. Instrum.*, 1970, **41**, 909-916.
8   H. Kubitschek, *Nature*, 1958, **192**, 234-235.
9   O. Saleh, L. Sohn, *Rev. Sci. Instrum.*, 2001, **72**, 4449-4451.
10 O. Saleh, L. Sohn, *Nano Lett.*, 2003, **3**, 37-38.
11 O. Saleh, L. Sohn, *Proc. Natl. Acad. Sci.,* 2003, **100**, 820-824.
12 P. Mitchell, et al., *Developmental Bio.*, 2005, **293**, 240-252.
13 A. Irintchev, et al., *Developmental Dynamics.*, 1994, **199**, 326-337.
14 U. Kaufmann, et al., *Cell Tissue Res.*, 1999, **296**, 191-198.
15 R. Moore and F. Walsh, *Development*, 1993, 117, 1409-1420.
16 I. Conboy, et al., *Science*, 2003, **302**, 1575-1577.
17 W. Dixon, *Ann. Math. Stat.*, 1950, **21**, 488-506.
18 W. Dixon, *Ann. Math. Stat.*, 1951, **22**, 68-78.
19 D. Rorabacher, *Anal. Chem.*, 1991, **83**, 139-146.
20 R. Sherwood, et al., *Cell*, 2004, **119**, 543-554.

COMBINING FLUORESCENCE IMAGING AND IONIC CURRENT DETECTION IN NANOCAPILLARIES

*V. V. Thacker**, S. M. Hernandez-Ainsa, J. Gornall*, L. J. Steinbock* and U. F. Keyser*
*Cavendish Laboratory, University of Cambridge, CB3 0HE Cambridge, United Kingdom

1 INTRODUCTION

Over the past decade ionic current detection of small molecules by nanopores has generated great interest for DNA sequencing and other biosensing applications[1],[2]. Using commercial laser pullers it is possible to produce nanocapillaries that are cheap and easy to fabricate[3],[4] . Our lab has recently demonstrated the ability to detect DNA folding with nanocapillaries[5]. We have now extended this system by combining ionic current detection with single-molecule fluorescence imaging. We present here recent data showing simultaneous imaging and ionic current detection of DNA fluorescently labeled with SYTOX Orange. Using a fast Electron Multiplying CCD (EMCCD) camera, we have verified the link between ionic current translocation events and the movement of the DNA into the nanocapillary. Further work will focus on elucidating the nature of so-called 'folding' DNA events as well as studies on the balance of electrophoresis and electroosmosis in nanocapillaries.

2 METHOD AND RESULTS

2.1 Nanocapillaries

Glass capillaries of between 20-40 nm in diameter are pulled using a commercial laser puller (P-2000B, Sutter). Details can be found in recent publications[5],[6]. Single nanocapillaries are then sealed into a custom designed microfluidic PDMS cell connecting the two reservoirs (Figure 1). DNA is driven through the capillary by an applied voltage and ionic current is measured using Ag/AgCl electrodes and a HEKA amplifier (EPC 800).

2.2 Fluorescent labeling of λ DNA

λ DNA (Fermentas, 0.5mg/ml) is diluted 100 fold to a final concentration of 500 ng/ml in 100 mM KCl, 1x TE buffer (Sigma). The DNA is then incubated in a fridge for 30 minutes with 500 nM SYTOX Orange (Invitrogen). SYTOX Orange is an intercalating dye and has the largest fluorescent enhancement upon binding for dyes excited around 530-540 nm[7], as well as a large secondary binding constant of 2.1 μM in 1xTE buffer which reduces the quenching of fluorescence. The protocol followed is identical to that given in[7]. At high salt concentrations (500mM or greater), the binding efficiency of the dye is reduced and the dye binds strongly to the nanocapillary[8].

2.3 Optical setup for imaging

The optical setup used is a subset of a larger microscope being built to combine fluorescent detection with Surface enhanced Raman Scattering (SERS) detection. In essence, light from a 532 nm laser is directed via a 532 nm dichroic mirror (DM 532, Semrock) and an objective (Olympus ULSAPPO 60x) onto the sample stage. Emitted fluorescence is collected via the same objective, passes through the 532 nm dichroic and is directed via a 633 nm dichroic mirror (DM 633, Semrock) to the sensor of a fast EMCCD camera (Andor iXon 860). The camera has a small chip size (128x128 pixels, 16 μm pixel size), and is designed for fast imaging with a maximum full frame rate of about 510 fps.

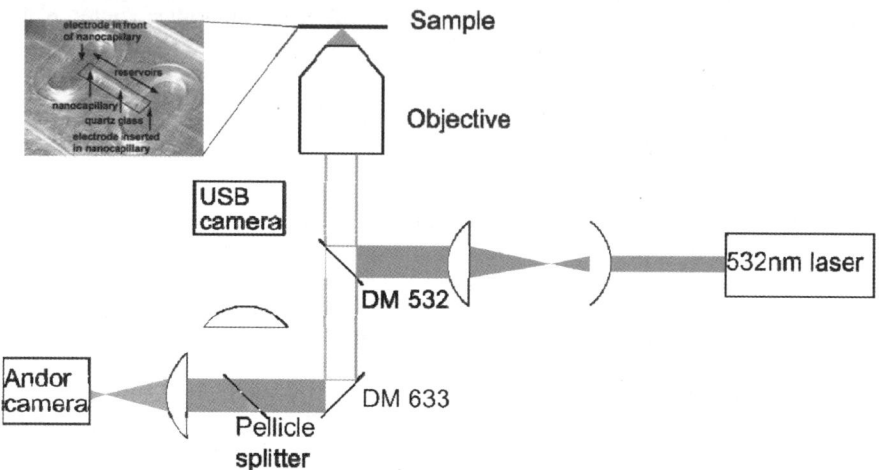

Figure 1: Optical setup for fluorescence detection. The USB camera enables additional bright field imaging.

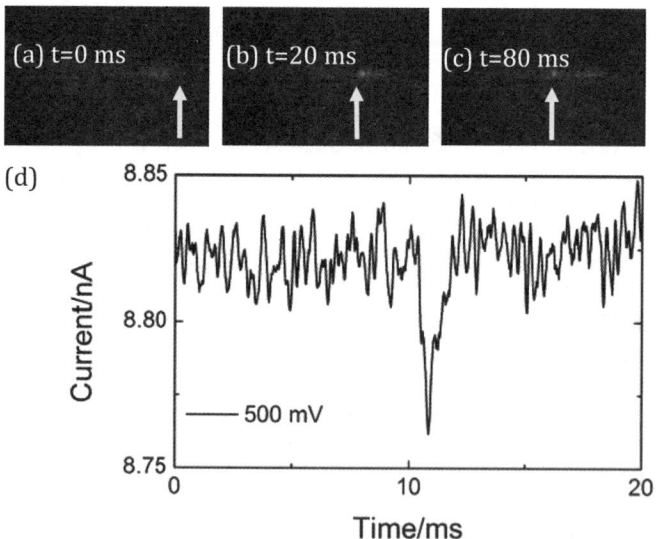

Figure 2: Three images from a video showing repeated DNA translocations. (a) A single λ DNA molecule diffuses close to the nanocapillary. (b) Molecule translocates through the nanocapillary. (c) The molecule is in the nanocapillary post translocation. (d) Accompanying current trace (filtered with a moving average for clarity).

2.4 Simultaneous detection of DNA translocation using fluorescence imaging and Ionic current detection

Using custom written software in LabVIEW, we simultaneously acquire optical and current readouts. A potential difference of 500 mV drives the fluorescently labeled DNA into the nanocapillary. Typical translocation times for such DNA are about 1 ms which compares well with previous experiments[5], but makes the imaging of the transport process challenging. Nevertheless we are able to observe the quick passage of single DNA molecules in the nanocapillary post translocation. This is demonstrated in Figure 2 below.

3 CONCLUSIONS

Here, we report first data on simultaneous fluorescence and ionic current detection of DNA translocation into nanocapillaries. We are working on improving the binding efficiency of dye-DNA at high salt concentrations to enable better imaging by reducing artifacts of dye binding to surfaces. Eventually, this will be integrated with Surface enhanced Raman Scattering (SERS) for the label-free identification of translocating molecules[9] and optical tweezers for controlling single DNA transport[10,11].

References

1 C. Dekker, *Nature Nanotechnology*, 2007, **2**, 209 - 215
2 D. Branton et al, *Nature Biotechnology*, 2008, **26**, 1146 - 1153
3 K. Lieberman et al, *Appl. Phys. Lett.*, 1994, *65*, 648
4 A. Bruckbauer et al, *NanoLett.* 2004, **4**, 1859–1862
5 L. J. Steinbock et al, *Nano Letters*, 2010, **10(7)**, 2493 - 2497
6 L. J. Steinbock et al, *Biosensors & Bioelectronics*, 2009, **24(8)**, 2423-7
7 X. Yan et al, ***Anal Biochem.***, 2000, **286(1)**, 138-48
8 X Yan et al, *Anal. Chem.*, 2005, **77 (11)**, 3554–3562
9 S. Nie et al, *Science*, 21 February 1997, **275 (5303)**, 1102-1106
10 U. F. Keyser, *J. R. Soc. Interface*, 2011, **8 (63)**, 1369-1378
11 O. Otto et al, *Reviews of Scientific Instruments*, 2011, **82**, 086102

AN INTRODUCTION TO A NEW ION BEAM NANOPATTERNING INSTRUMENT AND ITS APPLICATION FOR AUTOMATIC WAFER SCALE NANOPORE DEVICE PRODUCTION:

N.L.Peto[1], A.Nadzeyka[1], S.Bauerdick[1] and J.Edel[2], T.Albrecht[2]

[1] Raith GmbH, Dortmund 44263, Germany
  mailto: Peto@Raith.de
[2] Department of Chemistry, Imperial College, London SW7 2AZ, UK
  mailto: joshua.edel@imperial.ac.uk

1 INTRODUCTION

Several different instrumental approaches (conventional SEM-FIB, TEM, HIM etc.) have all been used to produce a few usable nanopores for developing biological filtering and molecule screening applications. These techniques have produced nanopores with varying degrees of quality, precision, repeatability and success. Here we introduce a new class of Ion Beam Lithography (IBL) / nanofabrication instrument which overcomes many of the established instrumental issues, and for the first time permits the automated production of nanopore devices at the (4") wafer scale. Exact nanopore placement within an existing device design, and integration of the milling process with other techniques is straightforward. Initial results from nanopores with state of the art dimensions, quality and uniformity are shown, and the instrumental process is discussed.

2 METHOD AND RESULTS

2.1 ion beam milling of a Silicon Nitride membrane

Desirable nanopores must be small enough to permit single molecule transit for the target application, they must be round, have high aspect ratio, and minimal damage or implantation of contaminants around them. The *ionLiNE* system concept as well as the construction of the NanoFIB ion column, its performance and how it is optimised for nanopatterning and in particular for the production of nanopores is described

2.2 automated wafer level device production

In addition to milling high quality nanopores, automated production places additional constraints on instrumentation. Nanopores must be produced over an extended period of time, with high uniformity and with exact positioning within a pre-fabricated device. Step and repeat operation for wafer level processing is also required.
These factors require a highly stable platform such as the *ionLiNE* lithography instrument, with GDSII design based navigation and automated pattern recognition, an ion column

engineered for long term Gallium source emission and beam stability, excellent spot quality and selectivity and laser height sensing for working distance optimisation between devices.

The performance of the instrument with respect to each of these requirements is shown and discussed.

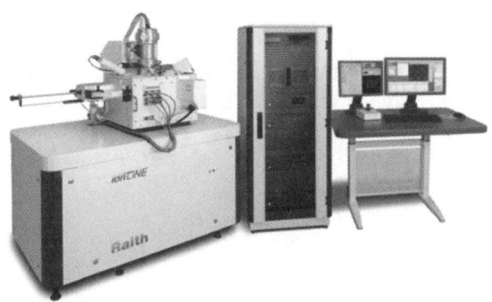

Figure 1*Raith ionLiNE^{TM}(Ion beam Lithography and NanoEngineering workstation)*

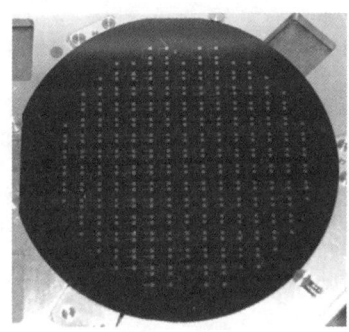

Figure 2 *A 4" NanoPore device wafer*

3 CONCLUSION

The automated production of exactly positioned, state-of-the-art nanopores, within integrated devices at the wafer scale is routinely achievable with this new IBL instrument. It is compatible and complimentary with other standard lithography and patterning techniques, for general application and production of different kinds of highly integrated devices, with reproducible parameters, and within reasonable time/costs.

References

1   B.Schiedt, L.Auvray, L.Bacri, G.Oukhaled, A.Madouri, E.Bourhis, G.Patriarche, J.Pelta, R.Jede, J.Gierak, Direct FIB fabrication and integration of ''single nanopore devices'' for the manipulation of macromolecules', Microelectronic Engineering 87, 2010

PH-REVERSED IONIC CURRENT RECTIFICATION DISPLAYED BY CONICALLY SHAPED NANOPORE WITHOUT ANY MODIFICATION

Jai Hai Wang and Er Kang Wang

State key Laboratory of Electroanalytical Chemistry, Changchun Institute of Applied Chemistry, Chinese Academy of Science, China.

1 INTRODUCTION

Ion current through nascent nanopore with conically shaped geometry in PET (polyethylene terephthalate) membrane sandwiched between two same buffer solutions at pH≤3 was routinely considered to exhibit no rectification if had, much weaker rectification than that for nanopore with negative surface charge, since the surface charge on the membrane decrease to zero along with decreasing the pH value of the buffer solution down to the pKa of carboxylic acid. However, in this study, we discovered that in the buffer solution with low ionic strength at pH values below 3, the conically shaped nanopores exhibited much distinct ion current rectification as expected for nanopores with positive surface charge, if voltages beyond ±2v range were scanned. We reasoned that the current rectification engendered by the positive surface charge of conical nanopore was due to further protonation of the hydrogen bonded hydrogel layer or neutral carboxylic acid inside the nanochannel. Therefore, our results enrich the knowledge about nanopore technology and indicate that nanofluidic diode based on pH-reversed ion current rectification through conical nanopore can be achieved without any modification of the PET membrane.

2 METHOD AND RESULTS

2.1 Nanopore Fabrication

Nanopores with the tip diameter of 30 nm was fabricated with surfactant-protected one-step etching method described by Ali for preparing conically shaped nanopores in tracked poly(ethylene terphthalate membranes).[1] Protecting solution (6 M NaOH + 0.07% 2A1) was placed on one side of the membrane, 6 M sodium hydroxide solution was placed on the UV-treated side to control the diameter of big opening of the nanopore. The whole etching process was carried out at 40 °C. When desired current was reached, 1 M HCl solution was placed on both side of the membrane to stop the etching process. The other procedures were same as those for the two-step etching method[2].

2.2 Electrochemical Assessment of Nanopore

We fabricated asymmetrical conical nanopore with 30 nm diameter of narrow opening by surfactant-protected one–step etching method, which has been claimed to produce better quality of conically shaped nanopore.[1] Indeed, we observed significant ion current rectification of the nascent nanopore. It has to be noted that surfactant 2A1 has been used in the whole etching process. As shown in Fig. 1(a) (black curve), without washing the nascent nanopore with ethanol solution, addition of aqueous HCl solution at pH 2 did not produce rectification properties as expected for nanopore with positive surface charge. In contrast, it produced ion current rectification as expected for nanopore with negative surface charge. Considering that it was highly possible that surfactant 2A1 having two negative charges can be adsorbed onto the polymer surface, we reasoned that this kind of ion current rectification was due to negative charges of surfactant 2A1 which has not been fully removed even after washing with plenty of water. Therefore, we immersed the membrane in pure ethanol solution overnight. Measurement of I-V curve showed that at aqueous HCl solution at pH 2, the nanopore after removal of surfactant 2A1 significantly rectified the current in the direction as expected for nanopore with positive charge (Fig. 1(a), grey curve). The ion rectification ration can reach 53 at absolute voltage value of 6 V as shown in Fig. 1(b) (grey curve). So, the reversal of surface charge in the nanopore wall indeed happened after the surface-adsorbed surfactant with negative charge had been eliminated.

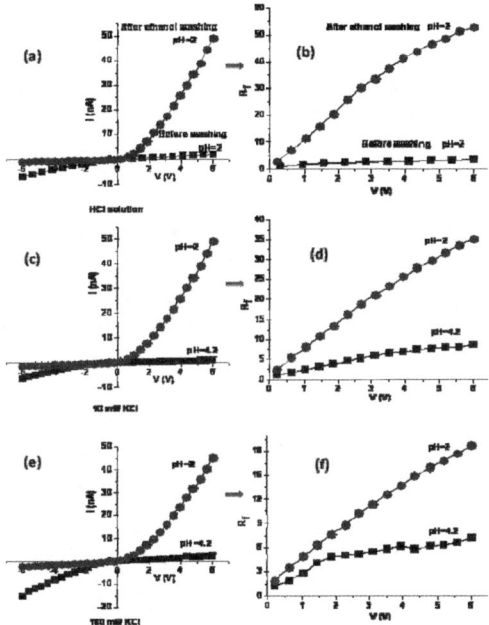

Figure 1. Current-voltage curves of the nanopore fabricated from surfactant-protected one-step etching method and corresponding plot of voltage-dependent ion current rectification. (a) Current-voltage curve measured in aqueous HCl solution at pH 2 before ethanol washing (square) and after ethanol washing (circle), (c) Current-voltage curve measured in 10 mM KCl solutions at pH 2 (circle) and at pH 4.2 (square), (e) Current-voltage curves measured in 100 mM KCl solutions at pH 2 (circle) and at pH 4.2 (square), (b, d and f) corresponding plots of voltage-dependent ion current rectification ratio. All the solutions were adjusted to the desired pH values by concentrated HCl. The diameters of narrow opening and big opening of the conical nanopore are 30 nm and 200 nm, respectively.

3 CONCLUSION

In summary, this study has demonstrated that asymmetrically conical-shaped nanopore in PET membrane had capability of switching the surface charge polarity from negative to positive , leading to formation of nanofluidic diode without any modification and the existence of heavy metal on. In order to observe this interesting phenomenon, proper ion strength and low pH value have to be used. We believe our study enriches the knowledge about nanopore technology which has attracted significant attention in the past years. Our result complements the previous studies which traditionally considered that conically shaped nanopore in PET membrane sandwiched between two buffer solutions at pH≤3 exhibited no rectification, if have, much weaker rectification than that for negatively charged surface, since the surface charge on the membrane was considered to approach zero along with decreasing the pH value of the buffer solution down to the pKa of carboxylic acid. Nevertheless, in this study, we discovered that in the buffer solution at low pH values and with proper ionic strength, the conically shaped nanopores with small diameter of narrow opening exhibited distinct ion current rectification as expected for nanopore with positive surface charge. We reasoned that the current rectification engendered by the positive charge of the surface of conical nanopore was due to further protonation of the neutralized carboxylic acid or hydrogen bonded gel-layer inside the nanopore. Therefore, pH-reversed ionic-current rectification via conical nanopore can be achieved without any modification of the PET membrane, facilitating the easy fabrication of nanofluidic diode in the later applications.

References

1   M. Ali, V. Bayer, B. Schiedt, R. Neumann and W. Ensinger, *Nanotechnol.*, 2008, **19**, 485711-485719.

2   J. E. Wharton, P. Jin, L. T. Sexton, L. P. Horne, S. A. Sherrill, W. K. Mino and C. R. Martin, *Small*, 2007, **3**, 1424-1430.

MODELLING SOLID-STATE NANOPORES WITH A COMBINATION OF THE POISSON-NERNST-PLANCK EQUATIONS AND BROWNIAN DYNAMICS

L. van Oeffelen[1,2], W. Van Roy[1], D. Charlier[2], L. Lagae[1,3] and G. Borghs[1,3]

[1] IMEC, Kapeldreef 75, B-3001 Leuven, Belgium
[2] Microbiology, Vrije Universiteit Brussel, Pleinlaan 2, B-1050 Brussel, Belgium
[3] Department of Physics, Katholieke Universiteit Leuven, Celestijnenlaan 200D, 3001 Heverlee, Belgium

1 INTRODUCTION

In principle, different physical models can be applied to simulate ion currents through nanopores. These include the continuum Poisson-Nernst-Planck (PNP) equations, Brownian dynamics (BD) and molecular dynamics (MD). In practice however, it is infeasible to perform molecular dynamics simulations long enough to simulate a sufficient amount of ion translocation events to determine a current, and even Brownian dynamics is often computationally too demanding to model the access resistance of a nanopore in a realistic way. As a consequence, nanopores are generally modeled using the PNP equations, even though this model does not always accurately describe the ionic movement within the nanopore itself, as ions are not treated as discrete entities and water is regarded as a static continuum[1]. The former issue can be solved by both Brownian and molecular dynamics, the latter only by molecular dynamics. Therefore, realistic simulations of ionic currents may be feasible if the PNP equations in the reservoirs are combined with Brownian and/or molecular dynamics within the nanopore. Here, we will focus on the combination of PNP with Brownian dynamics.

As Brownian dynamics is inherently time-dependent, the PNP-BD combination requires a time-dependent PNP simulator instead of the usual steady-state formulation. In this paper, we will first explain our proposed route to combine PNP and Brownian dynamics. Then, as a first step, we will present time-dependent PNP simulations of nanopores with different geometries and surface charges, and show that these simulations yield valuable new insights.

2 METHOD AND RESULTS

2.1 Combining PNP and BD

To combine the PNP equations with Brownian dynamics, we use the rate theory formulation of Brownian dynamics[2], in which drift-diffusion results from random hopping over energy barriers. The Nernst-Planck (drift-diffusion) equation can be derived from the same theory by calculating ionic fluxes as expected values: instead of discrete ions performing a random walk governed by field-dependent rate constants, ion concentrations are displaced according to these rate constants. Hence, a combined simulation can be obtained by meshing the complete simulation domain, thereby creating elementary volumes, and defining the domain where rate theory is applied as the space where the elementary volumes are so small that they are expected to contain less than e.g. 10 ions. The remainder of the simulation domain is then modeled with the Nernst-Planck equation and the time-dependent continuity equation, while the Poisson equation is solved over the complete domain. At the boundary between the domains, discrete ions are passed, changing the concentration at one side and the number of ions that are tracked by Brownian dynamics at the other side.

2.2 Geometry and boundary conditions

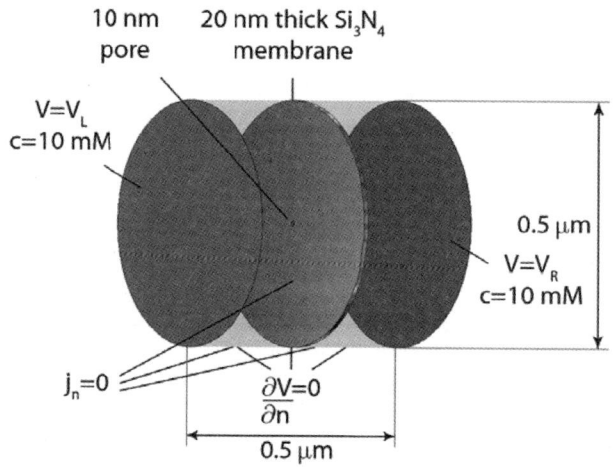

Figure 1 *Geometry and boundary conditions of the simulations.*

As model systems, we consider cylindrical and conical nanopores with a 10 nm inner diameter in a 20 nm thick Si_3N_4 membrane. The nanopore is immersed in a 10 mM KCl solution, and modeled as depicted in Figure 1. For convenience, we use equal diffusion coefficients for K^+ and Cl^-: $D(K^+)=D(Cl^-)=2\times10^{-9}m^2/s$. As boundary conditions, we apply voltages and concentrations at the left and the right of the simulation domain, and set the ionic fluxes through the remaining boundaries of the liquid equal to zero, as well as the derivative of the voltage normal to the remaining outer boundaries of the simulation domain. In contrast with earlier reports[3,4,5], these boundary conditions are less restrictive: they allow for variations of concentrations within the solutions contacting the nanopore, and a non-zero normal electric field component on the pore wall within the membrane.

2.3 Time-dependent versus steady state

Figure 2 shows the currents carried by the K^+ and the Cl^- ions 0.2 μs after applying a voltage step of $V_L-V_R = 0.1$ V across a membrane containing an uncharged cylindrical pore. Although the net ionic current is spatially uniform (the capacitance associated with the membrane is completely charged at this point), the individual ionic contributions are not. This can be understood by the fact that at each side of the membrane, one ionic species is drawn towards the membrane while the other is removed from it. This creates a region where both ionic species are depleted, and induces co-diffusion of K^+ and Cl^- towards the membrane. This process would take about 30 μs to reach the boundaries of the simulation domain (250 nm), and about an hour in a realistic situation with mm sized compartments. Hence, true steady state cannot be reached, even though the current may level off quite rapidly during a measurement or in a simulation. We used steady-state simulations to assess whether the time-dependent simulations have converged sufficiently to be representative for the measured current. In this case, we found that there is no detectable change in current after 0.2 μs.

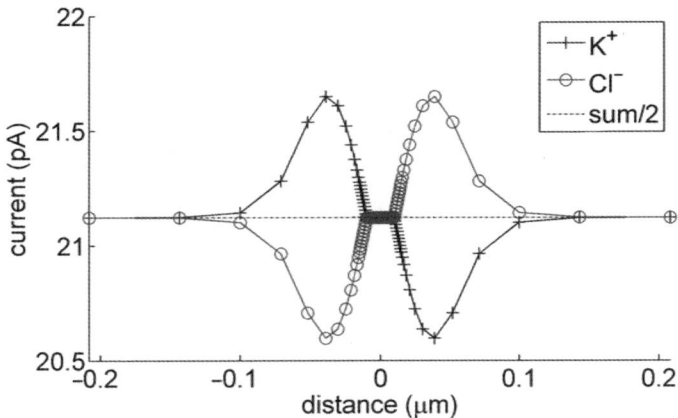

Figure 2 *Currents carried by the K⁺ and the Cl⁻ ions 0.2 μs after applying a voltage step of V_L-V_R = 0.1 V across an uncharged membrane with a cylindrical nanopore. The plotted values are the currents integrated over planes parallel with the membrane.*

2.4 Influence of surface charges: ion concentration polarization causes conductance changes and ion current rectification

In a next step, we study the influence of surface charges of −50 mC/m² appearing in the configurations shown in Figure 3. For this purpose, we initialize our time-simulations with the steady-state situation when 0 V is applied across the membrane, i.e., after the electric double layer has been built up. Figure 4 shows the currents 0.2 μs after applying a voltage step of V_L-V_R = 0.1 V across the membrane with configuration (a). We observe virtually perfect ion selectivity within the nanopore but not within the bulk, in agreement with an almost complete depletion of Cl⁻ ions inside the nanopore. This causes ion concentration polarization[6], also called the ion-enrichment and ion-depletion effect[7]: both ion concentrations decrease at the left and increase at the right, giving rise to co-diffusion of both ionic species respectively towards and away from the membrane. As conductivities are proportional to concentrations, the conductivity at the left decreases, while it increases at the right. The net effect is a conductance decrease, and a slower convergence towards the steady-state current as compared to the uncharged nanopore: after 0.2 μs, the current still decreases with about 4%.

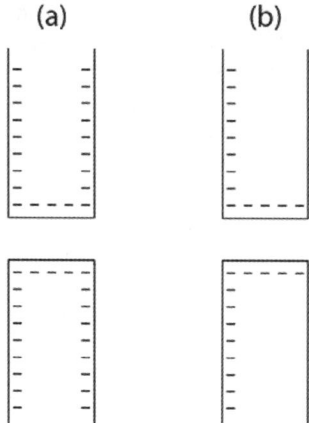

Figure 3 *Simulated surface charge configurations.*

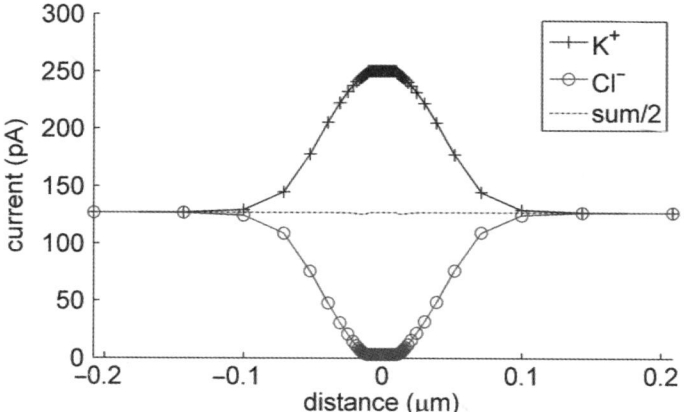

Figure 4 *Currents carried by the K^+ and the Cl^- ions 0.2 μs after applying a voltage step of V_L-V_R = 0.1 V across the membrane in Figure 3(a) with a surface charge of –50 mC/m².*

With the asymmetric configuration (b), ion current rectification is observed as shown in Figure 5. For positive voltages, concentration polarization occurs as a decrease of concentrations at the left of the nanopore, where the conductivity is high due to the accumulation of K^+ ions, and an increase of concentrations within and at the right of the pore where the conductivity is low, as shown in Figure 6. This results in a net positive effect on the conductance, which increases with increasing voltages. The opposite reasoning holds for negative voltages (Fig. 7). Hence, ion current rectification can be explained by concentration polarization in combination with a built-in asymmetry. In contrast with the simulations reported by Vlassiouk et al.[4], where I-V curves present a kink, the shape of the I-V curves after 0.2 μs and at steady state shown here corresponds well to the shape of experimental I-V curves[8]. This is most likely due to the difference in boundary conditions used for the Poisson equation.

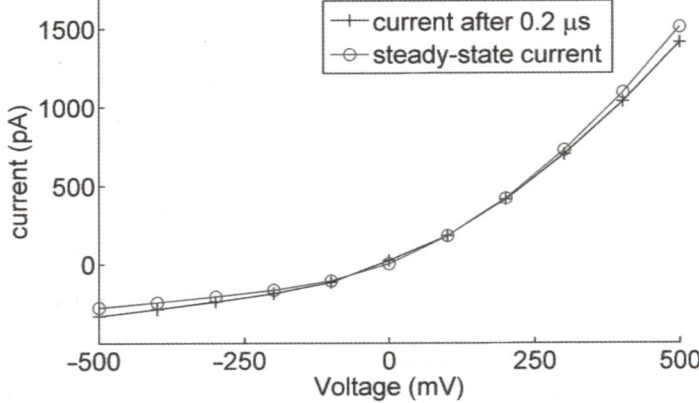

Figure 5 *Current after 0.2 μs and at steady state versus applied voltage for the membrane in Figure 3(b).*

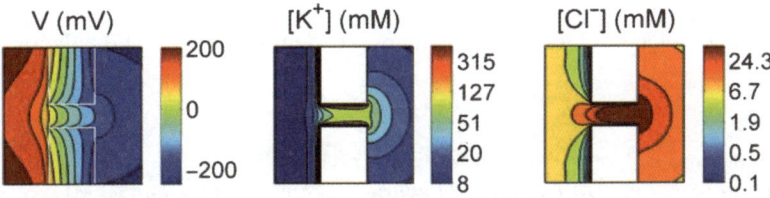

Figure 6 *Voltage and concentration profiles 0.2 μs after applying V_L-V_R = +0.5 V across the membrane in Figure 3(b). The values are plotted for a 60 nm by 60 nm area through the pore axis.*

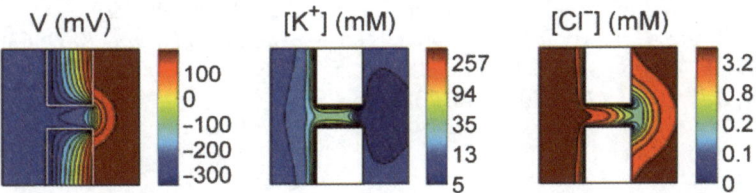

Figure 7 *Voltage and concentration profiles 0.2 μs after applying V_L-V_R = −0.5 V across the membrane in Figure 3(b). The values are plotted for a 60 nm by 60 nm area through the pore axis.*

2.4 Nanopore shape effects: current rectification in conical nanopores

A conical nanopore was simulated with uniform surface charges of −50 mC/m^2 and an opening angle of 10°, resulting in a largest opening with a diameter of 13.5 nm. Although rectification has previously been reported for track-etched conical nanopores[9], this effect was negligible in our simulation. We expect that this lack of rectification is due to the limited length of the simulated nanopore (20 nm) compared to track-etched membranes (12 μm), which results in a reduced asymmetry. In membranes with a thickness of several μm, we expect that ion concentration polarization mainly affects concentrations outside the nanopore at the entrance of the smallest pore opening, and within the nanopore. The fact that the nanopore conductance is mainly sensitive to the concentration within the pore then explains the rectification effect.

3 CONCLUSION

Realistic simulations of ionic currents through nanopores may be feasible if the PNP equations are combined with Brownian and/or molecular dynamics within the nanopore. Such a combination requires a time-dependent PNP implementation. Simulations obtained with the time-dependent PNP model alone already yield new insights: 1) although electrical currents quickly converge to a steady value after changing the applied voltage, steady state is generally not reached for the ionic concentrations, due to slow co-diffusion processes, 2) ion concentration polarization can have an effect on the conductance of a charged nanopore, and the current convergence, and 3) concentration polarization in combination with built-in asymmetries is responsible for ion current rectification.

4 ACKNOWLEDGEMENTS

Liesbeth van Oeffelen has a PhD fellowship of the Flemish Research Foundation.

References

1   S.H. Chung and B. Corry, *Soft Matter*, 2005,**1**,417-427.
2   B. Hille, *Ion Channels of Excitable Membranes*, Sinauer associates, 2001.
3   I. Vlassiouk, S. Smirnov and Z. Siwy, *Nano Letters*, 2008, **8**, 1978-1985.
4   I. Vlassiouk, S. Smirnov and Z. Siwy, *ACS Nano*, 2008, **2**, 1589-1602.
5   J. Cervera, B. Schiedt, R. Neumann, S. Mafé and P. Ramírez, *J. Chem. Phy.*, 2006, **124**, 104706.
6   S. J. Kim, Y.A. Song and J. Han, *Chem. Soc. Rev.*, 2010, **39**, 912-922
7   G. Li, S. Wang, C. K. Byun, X. Wang and S. Liu, *Anal. Chim. Acta*, 2009, **650**, 214-20.
8   I. Vlassiouk, Z. Siwy, *Nano Letters*, 2007, **7**, 552-556.
9   Z. Siwy, *Adv. Func. Mat.*, 2006, **16**, 735-746.

BIOMIMETIC NANOPORES WITH AMPHOTERIC AMINO ACID GROUPS. EFFECTS OF A pH GRADIENT ON THE IONIC CONDUCTANCE AND SELECTIVITY

A. Alcaraz[1], M. Ali[2], W. Ensinger[2], S. Mafe[3], F. Münch[2], S. Nasir[2] and P. Ramirez[4],*

[1]Dept. de Fisica, Universitat Jaume I de Castello, Spain.
[2]Dept. of Material- and Geo-Sciences, Technische Univ. Darmstadt, Germany.
[3]Dept. de Fisica de la Terra y Termodinamica, Universitat de Valencia, Burjassot, Spain.
[4]Dept. de Fisica Aplicada, Universitat Politecnica de Valencia, Spain.

1 INTRODUCTION

Biological ion channels have been an inspirational source in the development of new devices in a variety of biotechnological and analytical applications. During the last years, synthetic nanopores have been carefully engineered to mimic distinctive features of biological ion channels. Conductance in the pS–nS range, pH-dependent ion selectivity, fluctuations of current between open and closed states, flux inhibition caused by protons or divalent cations, current rectification, among others are found both in synthetic and biological channels. Thus, the comprehension of the mechanisms by which biological channels regulate the transport of ions and the electric signal transduction at the molecular level is crucial for building ionic circuits in the growing field of nanofluidics. We show here that gold coated polymeric nanopores modified with amino acid (Cysteine) chains are able to imitate the electrical responses of the OmpF porin, a wide channel found in the external membrane of the bacteria Escherichia Coli. As a consequence of the amphoteric nature of the amino acid chains anchored to pore wall, both the electrical conductance and the ionic selectivity are strongly dependent on pH, just as it is the case of ion channels. Under symmetrical pH conditions, both pores display ohmic conduction and well-defined conductance levels. However, when a pH gradient is applied, the distributions of fixed charges along the pores become asymmetric and a diode-like performance is found in both systems. In the case of a basiclacidic pH gradient, the fixed charge distribution in the pores resembles to that of solid-state p-n junctions and bipolar membranes. This means that the conductive properties of a tunable nanostructure can be manipulated at will by changing the pH of the external solutions, inspiring potential applications such as uphill ionic transport in presence of a proton gradient, energy storage membranes based on ion pumping, and miniaturization of pH-based nanofluidic sensors and actuators.

2 METHOD AND RESULTS

The OmpF is composed of three identical monomers. Each monomer opens an aqueous pore that communicates the inner and outer parts of the cell. At physiological pH conditions, the channel contains positive and negative residues in the protein matrix. The experiments were conducted on single ion channels reconstituted on planar lipid bilayer membranes [01]. The single cylindrical, biomimetic nanopore was fabricated in a polyethylene terephthalate (PET) membrane using the track-etching method. The electroless deposition of a thin Au layer on the surface and inner walls, and the chemisorption of Cysteine-monolayer on the Au layer were achieved using the methods described in [02, 03, 04]. The OmpF ion chanel reconstituted on planar lipid bilayer and the biomimetic nanopore were mounted between the two halves of a conductivity cells. An Ag/AgCl electrode was placed into each half-cell solution and a picoammeter/voltage source was used to measure the *I-V* curves. In all measurements the OmpF channel and the bio-mimetic nanopore separated by two 0.1 M KCl aqueous solutions at room temperature.

Figure 1 shows the experimental *I-V* curves of the OmpF channel (a) and the Au-Cysteine modified nanopore (b) for acidic, neutral, and basic pH values under symmetrical conditions (the same pH value and KCl concentration in the two external solutions). The experimental data reveal that the two pores exhibit resistor-like, ohmic behavior under symmetrical pH conditions. The OmpF channel contains positive and negative residues that are in charged state at neutral pH. Since the concentration of negative residues is higher than that of the positive ones, the channel exhibits a slight selectivity to cations. When the pH of the surrounding solutions is decreased to acidic values, the negative residues become neutral, while the positive ones remain charged. As a result, the channel selectivity is reversed and the pore becomes anion-selective. When the pH of the solutions is increased to basic values, these residues become neutral while the negative ones are charged. The result is a net negative fixed charge density at the pore walls. In the case of the *I-V* curves of the synthetic nanopore (b). The three conductance levels correspond to the Cysteine groups in positive form (acidic pH values), in neutral, zwitterionic form (intermediate pH values), and in negative form (basic pH values). Contrary to the case of the OmpF pore, the slopes of the *I-V* curves at low and high pH values are similar, in agreement with the fact that each Cysteine group has one positive and one negative residue. At intermediate pH, the net fixed charge concentration is zero and the conductance of the pore decreases, as indicated by the low slope of the *I-V* curve.

Figure 2 shows the *I-V* curves of the OmpF channel (a) and the synthetic nanopore (b) under a pH gradient applied to the external solutions. The experiments correspond to acidic‖neutral, basic‖neutral, and basic‖acidic pH gradients. The *I-V* curves of the two systems deviate now from linearity and display a non-ohmic, diode-like behavior showing rectification, with well-defined *on* (high conducting) and *off* (low conducting) states. The *I-V* curves can be interpreted as follows: the pH gradient applied leads to asymmetric fixed charge distributions (see the cartoons close each curve) along the pore walls. As a result, the resistances at the entrances of the OmpF channel and the nanopore dictate the experimental *I-V* curves: higher currents are obtained when the ions enter the pore end with the higher opposite charge (see the cartoons of the *I-V* curve). Interestingly, the basic‖acidic pH configuration produces asymmetric, bipolar fixed charge distributions analogue to those found in solid-state p-n junctions, conical nanopores with alternating positive and negative fixed charge regions, and bipolar ion-exchange membranes [05,06]. The rectification ratios, defined as the absolute value of the ratio between the electric current in the on and off conductive states, are low in the case of the acidic‖neutral and basic‖neutral configurations. Our experiments show that, in the case of the OmpF channel, these ratios can be increased by applying the a basic‖acidic pH, yielding a bipolar charge distribution. The basic‖acidic pH configuration also leads to a bipolar distribution of fixed charges in the nanopore, but the increase in the rectification ratio observed is not so pronounced, probably because the nanopore is very long compared with the ion channel and the fixed charge distribution becomes neutral in most of the inner part of the pore.

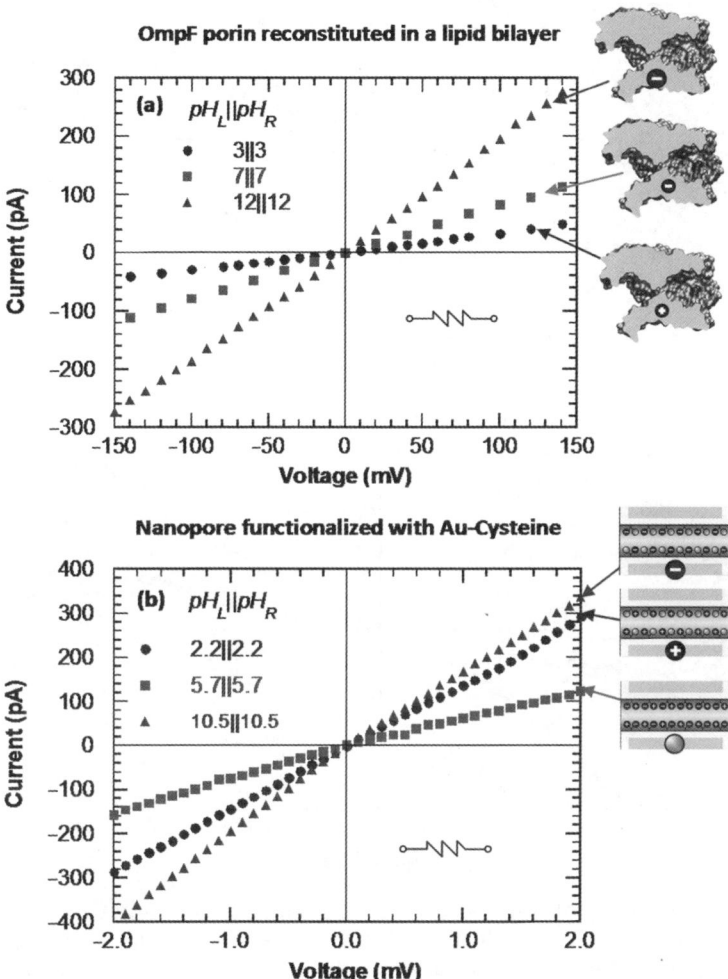

Figure 1: *I-V* curves of the OmpF channel and the bio-mimetic nanopore under symmetrical pH conditions. Both nanostructures behave as resistors and exhibit linear *I-V* curves with well-defined conductance levels that can be tuned by the pH of the external solutions. However, no rectification is observed in the range of pH values imposed. This fact suggests that the shape of the nanopore is approximately cylindrical and that the slight structural asymmetry of the protein channel has a negligible effect on the *I-V* curve. The experimental *I-V* characteristics are essentially controlled by the electrostatic interactions between the permeating ions and the ionizable residues within the porin, while the channel geometry remains mostly unaffected by the solution pH and the voltage.

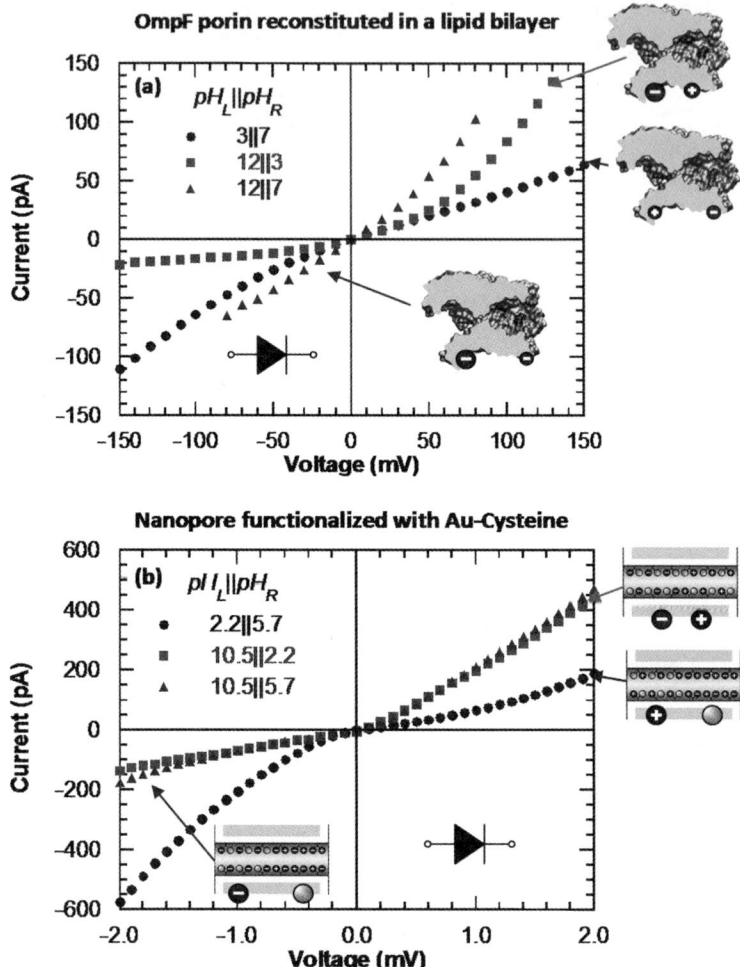

Figure 2: *I-V* curves of the OmpF channel and the bio-mimetic nanopore under asymmetrical pH conditions. When a pH gradient is applied, the distributions of fixed charges along the two pores become asymmetric and the resulting *I-V* curves show diode-like rectification, with well-defined on (high conductive) and off (low conductive) states. The characteristics of the *I-V* curves depend basically on the resistance experienced by the electric current when entering the pore openings. In particular, the on state appears when the mobile ions enter first the pore region of fixed charges with opposite sign.

3 CONCLUSION

The experiments reported here show analogue conductive properties for the biological ion channel (OmpF porin) and cylindrical nanopores functionalized with amphoteric Cysteine groups The strong pH dependence found in both *I-V* curves and ion selectivity can be rationalized paying attention to the reversible protonation/deprotonation of the titratable groups located on the pore walls. Under symmetrical pH conditions, both systems display linear *I-V* curves. This ohmic conductance can be tuned by the pH of the external solutions.

However, when a pH gradient is applied, the distributions of fixed charges along the two pores become asymmetric and the resulting *I-V* curves show diode-like rectification. The characteristics of the *I-V* curves depend essentially on the resistance found by the electric current in the pore openings. In the case of a basic/acidic pH configuration, the fixed charge distributions and the *I-V* curves are similar to those of solid-state p-n junctions and bipolar membranes [05]. It is important to note that the above rectification properties can be controlled externally in a reversible way by simply adjusting the pH of the external solutions. This fact opens the door to use a single reconfigurable nanostructure in practical applications such as uphill ionic transport in presence of a proton gradient [07, 08], energy storage membranes based on ion pumping [07, 09, 10], and pH-based nanofluidic sensors and actuators [11].

References

1   A. Alcaraz, P. Ramirez, E. Garcia-Gimenez, M. L. Lopez, A. Andrio, V. M. Aguilella, *J. Phys. Chem. B* 2006, **110**, 21205.
2   B. Yameen, M. Ali, R. Neumann, W. Ensinger, W. Knoll and O. Azzaroni, *Nano Lett.* 2009, **9**, 2788.
3   F. Muench, M. Oezaslan, T. Seidl, S. Lauterbach, P. Strasser, H. J. Kleebe and W. Ensinger, *Appl. Phys. A* 2011, **105**, 847
4   F. Muench, U. Kunz, C. Neetzel, S. Lauterbach, H. J. Kleebe and W. Ensinger, *Langmuir* 2011, **27**, 430.
5   P. Ramirez, H.-J. Rapp,S. Mafe and B. Bauer. *J. Electroanal. Chem.* 1994, **375**, 101
6   I. Vlassiouk and Z. S. Siwy, *Nano Lett.* 2007, **7**, 552.
7   P. Ramírez, A. Alcaraz and S. Mafe. J. *Membrane Sci.* 1997, **135**, 135
8   J. R. Ku, S. M. Lai, N. Ileri, P. Ramirez, S. Mafe and P. Stroeve, *J. Phys. Chem. C* 2007, **111**, 2965.
9   Z. Siwy, Y Gu, H. A. Spohr, D. Baur, A Wolf-Reber, R. Spohr, P. Apel and Y. E Korchev. *Europhys. Lett.* 2002, **60**, 349.
10 J. Cervera, P. Ramirez, S. Mafe and P. Stroeve. *Electrochim. Acta* 2011, **56**, 4504.
11 M. Ali, S. Mafe, P. Ramirez, R. Neumann and W. Ensinger, *Langmuir* 2009, **25**, 11993.

FORCES ON DNA IN CONFINEMENT AS MEASURED BY OPTICAL TWEEZERS

O. Otto[1], N. Laohakunakorn[1], L. J. Steinbock[1], U. F. Keyser[1]

[1] Cavendish Laboratory, University of Cambridge, JJ Thomson Ave, Cambridge CB3 0HE, UK

1 INTRODUCTION

A detailed understanding of the voltage-driven translocation process of DNA through small nanopores[1-4] is still elusive[5, 6]. To shed light on the translocation process for controlling electrophoretic translocation through nanopores[7], we have developed an optical tweezers system with high-speed and real-time position tracking using video technology[8] to control and measure the force on DNA in nanopores[9]. We combine our optical tweezers with nanocapillaries[9, 10], single molecule biosensors with similar characteristics to nanopores made in solid-state membranes[11, 12]. With the nanocapillary located perpendicular to the trapping laser our experiments are less affected by laser-induced heating[13]. We demonstrated simultaneous measurements of electrophoretic force and ionic-current on single DNA molecules inside the orifice of a nanocapillary[9]. This allows for full control of the molecules in nanocapillaries. Here, we describe our first results on multiple DNA molecules in confinement and demonstrate the nanocapillary diameter determines the force scaling when several DNA strands are simultaneously stalled in the orifice.

2 METHOD AND RESULTS

2.1 Optical Tweezers with Nanocapillaries

A schematic drawing of our optical tweezers setup is shown in Figure 1. Based on a custom-built inverted microscope, we form a static optical trap inside the microfluidic cell we apply a 5 W ytterbium fiber laser (YLM-5-LP, IPG Laser, Germany) at a wavelength of $\lambda=1064$ nm. A pair of lenses L1 and L2 expands the beam which is then reflected by a dichroic mirror into our water immersion objective (UPlanSApo/IR, 60x, Olympus, Japan). The position of the optical trap relative to our microfluidic cell is controlled by a xyz piezoelectric nanopositioning system (P-517.3, E-710.3, Physik Instrumente, Germany). For coarse control the nanopositioning system itself is mounted onto a custom-made micrometer stage. Illumination is either done using a standard white light source (DC-950 Fiber-Lite, Edmund Optics, USA), or following a novel optical fiber approach (100 W mercury arc lamp, LOTOriel, UK and 600 μm multimode silica fiber, NA0.39, Thorlabs, UK)[3]. We use a CCD camera (Imaging Source, DMK31AF03, Germany) for wide-field imaging, while high-speed speed tracking is carried out by a CMOS camera (MC1362,

Mikrotron, Germany) which is used for position tracking only. The setup allows for tracking of single colloids at up to 10,000 frames per second with 2 nm resolution in realtime[1]. For ionic current measurements a Faraday cage is mounted on top of the micrometer stage with an Axopatch 200B amplifier (Molecular Devices, USA) with standard Ag/AgCl electrodes in the microfluidic cell (Figure 1).

Fabrication of the nanocapillaries is described in detail elsewhere[12]. In brief, quartz glass capillaries are heated with a laser pipette puller (P2000B, Sutter) are heated until melting and separated in a single pulling step. Our protocol yields a pair of nanocapillaries with diameters down to 20 nm[14]. These are mounted onto the optical tweezers using a custom-made microfluidic cell shown in the inset of Figure 1.

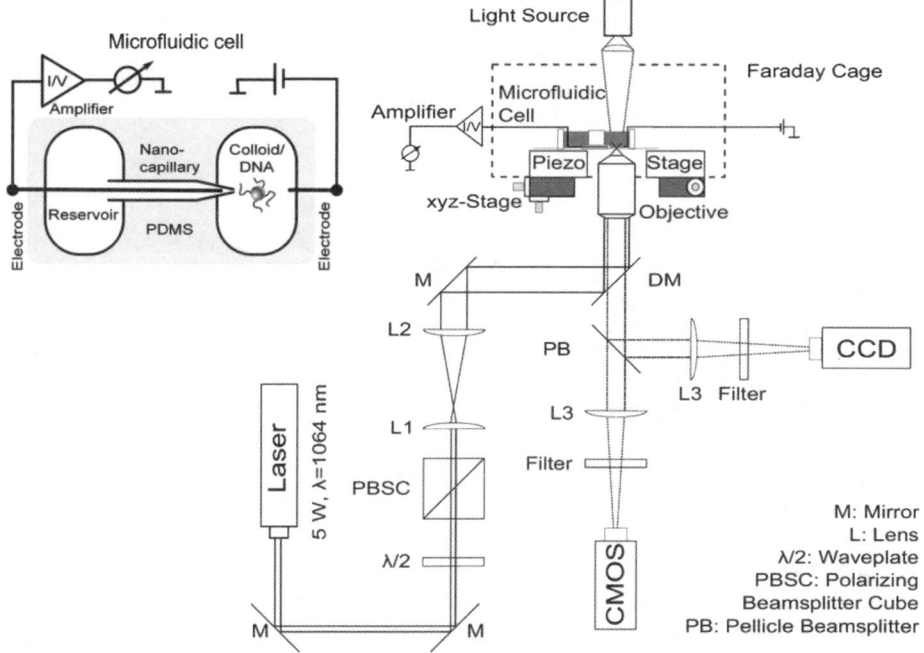

Figure 1 *Schematic drawing of optical tweezers setup and ionic current detection. The setup is based on a custom-built inverted microscope on an optical table. Light from a λ=1064 nm laser is expanded and coupled into a 60x water immersion objective overfilling the back aperture. The laser is focused inside a microfluidic cell which is illuminated from above using white light illumination (dashed line). A CCD camera is used to image the full field of view at 30 frames per second (fps) whereas a CMOS camera tracks the position of an optically trapped colloid at up to 10,000 fps.The upper left inset shows a schematic of our microfluidic cell containing the nanocapillary and the DNA coated colloid (not to scale).*

2.2 Force measurements on Multiple DNA strands

For single molecule experiments with DNA we use 2.1 μm PS colloids (Polysciences). We perform repeated power spectrum calibration for a number of particles to verify the proper grafting of λ-DNA on our polystyrene colloids. Empirically, the corner frequency of plain colloids is 20% higher than of coated ones carrying up to 40 DNA molecules. After calibration, a λ-DNA coated colloid is positioned in front of the capillary tip using the optical tweezers. Typically, the distance between tip and center of the colloid is approximately 3.5 μm. A constant positive potential will pull single DNA strands into the nanocapillary[3] due the negative backbone of the DNA. Molecules are stalled due to the force balance of electrophoretic force F_{ep}, Stokes friction force F_S, and restoring optical force F_O: $F_{ep} - F_S = F_O$. F_{ep} on the DNA is determined by the bare charge and the screening due to counter ions at the given salt concentration[5, 15]. The electro-osmotic fluid flow generated by the moving positive counter ions on the capillary and DNA surface lead to a counteracting F_S. Both forces result in an effective force which is balanced by F_O as shown in[6, 16]. DNA molecules attached to a single colloid were sequentially captured inside a 580 nm capillary (Figure 2a). The DNA capture force F_O was scaled by the applied potential and the distribution was analyzed. The histogram in Figure 2a shows five peaks corresponding to the number of DNA strands in the pore. The inset in Figure 2 demonstrates that the DNA molecules are not influencing each other as the force is proportional to the number of strands. In essence, we see no hydrodynamic interactions between DNA strands for this size of glass capillary.

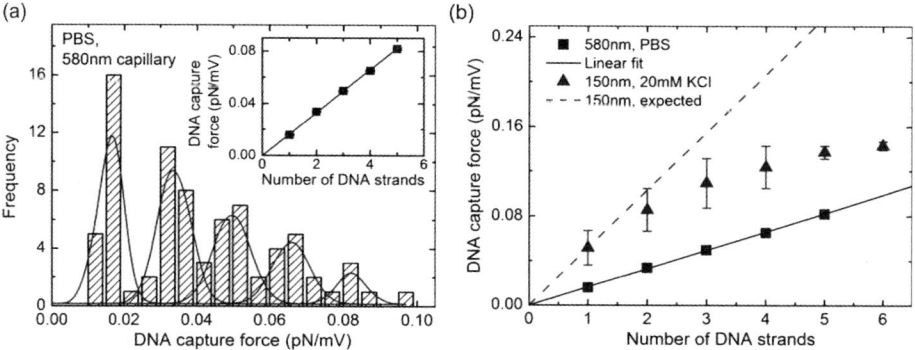

Figure 2 *(a) DNA confinement in a microcapillary. Analysis of multiple DNA capture events in a 580 nm borosilicate glass capillary. The graph summarises 29 measurements carried out with 4 colloids where multiple strands of DNA have been captured at a constant colloid-tip distance of D = 1.5 μm. For comparison the capture force of the DNA is scaled with the applied potential. The inset shows the mean scaled capture force per DNA strand taken from the Gaussian fits to the data. The fit reveals a linear dependency with a slope of 0.016 pN/mV per strand. The error is given by the symbol size. (b) Crowding effects in glass nanocapillaries. DNA capture force Is given as a function of the number of strands inside the capillary for experiments in PBS pH7.4 solution using a 580 nm borosilicate capillary (black squares, from (a)). Data from measurements in 20mM KCl, 2mM Tris pH8 solution using a 150 nm quartz nanocapillary are displayed as red triangles. The error is calculated from a series of measurements. The dashed red line is calculated for the 150 nm tip expecting a linear dependency between capture force and number of DNA strands inside the capillary.*

However, if the diameter of the capillary is reduced to 150 nm in diameter the DNA capture force decreases with the addition of each molecule to the nanopore (Figure 2b). In fact, the DNA capture force decreases significantly from ~0.06 pN/mV to less than ~0.01 pN/mV when 6 DNA strands are in the nanocapillary. In fact, starting from an initial value of 0.051 pN/mV for one strand, F_O raises to only 0.143 pN/mV after six molecules have been stalled (Figure 2b). This is only half the value expected from a linear relationship indicated by the dashed red line, calculated under the assumption that the force is additive. This is probably due to hydrodynamic interactions in addition with the confinement of the molecules in our nanocapillaries.

Our data shows that for 6 strands in the nanopore the DNA capture force is roughly 2.8 times larger than for a single molecule. This major discrepancy with the data for larger capillaries remains unexplained. Ghosal discussed that two DNA chains inside a nanopore could be modeled as a single molecule having the same ζ-potential but a larger effective radius, arguing that the same fractional area of the nanopore is blocked[16]. This model predicts that the force should increase much slower than with the square root of the number of molecules we observe (Figure 2b). Ghosal'a assumption should hold for nanopores with any diameter, which is also not the case as our data in larger nanocapillaries shows that the force is simply proportional to the number of molecules (Figure 2a). The reason remains unclear but is probably related to the hydrodynamic interactions which are dominating in smaller nanocapillaries.

3 CONCLUSION

We developed a setup which allows for the stalling of several DNA molecules simultaneously in nanocapillaries with diameters down to 150 nm. Our results demonstrate that the force on DNA molecules are linearly proportional to the number of molecules in the nanocapillary with diameters larger than 500 nm. In contrast, for nanocapillaries with diameters around 150 nm we observe that the force is not scaling linearly with the number of DNA strands but increasess with the square root. Our data is interesting for the modeling of DNA translocation in crowded environments suggesting that inter-molecular spacing is important for the electrophoretic forces. Quantitative modeling and more data for a range of salt concentrations in addition to measurements in smaller and larger nanocapillaries are needed to clarify the force on DNA strands in crowded environments. In the future, our setup is suitable to probe tension and relaxation dynamics of single DNA molecules inside and outside of nanocapillaries. We will also employ our system for the analysis of voltage-driven translocation through protein nanopores[17] or DNA origami nanopores[18].

References
1.   C. Dekker, Nature Nanotechnology **2** (4), 209-215 (2007).
2.   H. Bayley and C. R. Martin, Chem Rev **100** (7), 2575-2594 (2000).
3.   S. M. Bezrukov, Journal of Membrane Biology **174** (1), 1-13 (2000).
4.   J. J. Kasianowicz, E. Brandin, D. Branton and D. W. Deamer, Proceedings of the National Academy of Sciences of the United States of America **93** (24), 13770-13773 (1996).
5.   U. F. Keyser, B. N. Koeleman, S. Van Dorp, D. Krapf, R. M. M. Smeets, S. G. Lemay, N. H. Dekker and C. Dekker, Nature Physics **2** (7), 473-477 (2006).
6.   U. F. Keyser, S. van Dorp and S. G. Lemay, Chemical Society Reviews **39** (3), 939-947 (2010).
7.   U. F. Keyser, J R Soc Interface **8** (63), 1369-1378 (2011).
8.   O. Otto, J. L. Gornall, G. Stober, F. Czerwinski, R. Seidel and U. F. Keyser, J Optics-Uk **13** (4) (2011).
9.   O. Otto, L. J. Steinbock, D. W. Wong, J. L. Gornall and U. F. Keyser, Review of Scientific Instruments **82** (8) (2011).

10.   A. Bruckbauer, L. Ying, A. M. Rothery, D. Zhou, A. I. Shevchuk, C. Abell, Y. E. Korchev and D. Klenerman, J Am Chem Soc **124** (30), 8810-8811 (2002).

11.   J. Li, D. Stein, C. McMullan, D. Branton, M. J. Aziz and J. A. Golovchenko, Nature **412** (6843), 166-169 (2001).

12.   L. J. Steinbock, O. Otto, C. Chimerel, J. Gornall and U. F. Keyser, Nano Letters **10** (7), 2493-2497 (2010).

13.   L. J. Steinbock, O. Otto, D. R. Skarstam, S. Jahn, C. Chimerel, J. L. Gornall and U. F. Keyser, Journal of Physics-Condensed Matter **22** (45) (2010).

14.   L. J. Steinbock, A. Lucas, O. Otto and U. F. Keyser, Electrophoresis **in press** (2012).

15.   S. van Dorp, U. F. Keyser, N. H. Dekker, C. Dekker and S. G. Lemay, Nature Physics **5** (5), 347-351 (2009).

16.   S. Ghosal, Phys Rev E **74** (4) (2006).

17.   J. L. Gornall, K. R. Mahendran, O. J. Pambos, L. J. Steinbock, O. Otto, C. Chimerel, M. Winterhalter and U. F. Keyser, Nano Letters **11** (8), 3334-3340 (2011).

18.   N. A. Bell, C. R. Engst, M. Ablay, G. Divitini, C. Ducati, T. Liedl and U. F. Keyser, Nano Lett **12** (1), 512-517 (2012).

POTENTIAL DEPENDENCE OF DNA TRANSLOCATION

L. J. Steinbock, U. F. Keyser[1]

[1] Cavendish Laboratories, Cambridge University, JJ Thomson Avenue, CB3 0HE, Cambridge, UK

1 INTRODUCTION

Experiments using solid-state nanopores and biological nanopores have seen remarkable progress in the last decade.[1,2,3,4] Biological nanopores allow exact control of the atomic composition of the molecules through mutations.[5] Nevertheless current biological nanopores like α-haemolysin and *Mycobacterium smegmatis* porin A (MSPA) have inner diameters of 1.4 nm and 1 nm at their smallest constriction, respectively.[6,7] Therefore, only single-stranded DNA can pass through these channels which exclude translocating molecules of the size of dsDNA or protein-DNA complexes. On the other hand, solid-state nanopores are robust and have access to the large field of chip manufacturing techniques since they are made out of semiconductor materials.[8] Nevertheless, if the goal is to detect dsDNA molecules and their folding state with a cost-effective bench top technique glass nanocapillaries are an interesting alternative.[9,10]

Shortly after the first experiments using solid-state nanopores in silicon membranes, the effect of varying potentials on the translocation behavior was investigated.[2,8,11,12,13] Chen *et al.* presented results, that showed a linear dependence of the λ-DNA translocation time on the applied potential. This was followed up by Wanunu *et al.* demonstrating that the capture rate increases linearly with the concentration of the DNA.[12] Later Wanunu *et al.* revealed that for long DNA molecules such as λ-DNA the capture rate scales linearly with the applied potential.[13] This is due to the electric field attracting the DNA towards the nanopore, and hence increases the number of incoming DNA molecules when the potential is raised. The effect of different potentials on the amount of blocked charge was only recently studied in detail by Skinner *et al.* by translocating various nucleic acid polymers.[14]

2 METHODS AND RESULTS

This work will analyze the effect of potentials ranging from 0.3 V to 0.5 V on λ-DNA translocation through a nanocapillary immersed in 0.5 M KCl solution. The λ-DNA was diluted to a concentration, c_{dna}, of 1.03 nM and used to replace the solution in the reservoir in front of the nanocapillary. By applying a positive potential to the electrode inside the nanocapillary the λ-DNA was pulled through the nanocapillary causing decreases in the ionic current. We will investigate duration, current blockage and conductance change as a function of applied potential. Detailed materials and methods can be found in a previous publication.[15]After adding the λ-DNA to the reservoirs in front the nanocapillary positive potentials of 0.3, 0.4, and 0.5 V are applied to the Ag/AgCl electrode inside the nanocapillary. Typical current traces recorded at 0.4 V and 0.5 V can be seen in the left graph of figure 1(a) and 1(b). The current is zeroed by subtracting the baseline current which lies at 12.34±0.24 nA and 15.48±0.03nA for the experiment at 0.4 and 0.5 V. For each of these zeroed current traces, current$_z$, a histogram is generated, which shows up to two clear peaks (see right graph in figure 1a and 1b). This is expected and reported previously by the folding states of the DNA causing quantized current blockades.[15,16] For this analysis, only the peak caused by a single unfolded dsDNA strand is fitted with a Gauss function, whose centers are positioned at 23.73±0.09 pA and 31.93±0.13 pA for the experiment at 0.4 and 0.5V, respectively. These values, referred to as I_{un} for unfolded DNA, are visualized with dashed lines in the graphs of figure 1 and agree well with the current traces. As anticipated, the I_{un} value increases with higher potentials.[16]

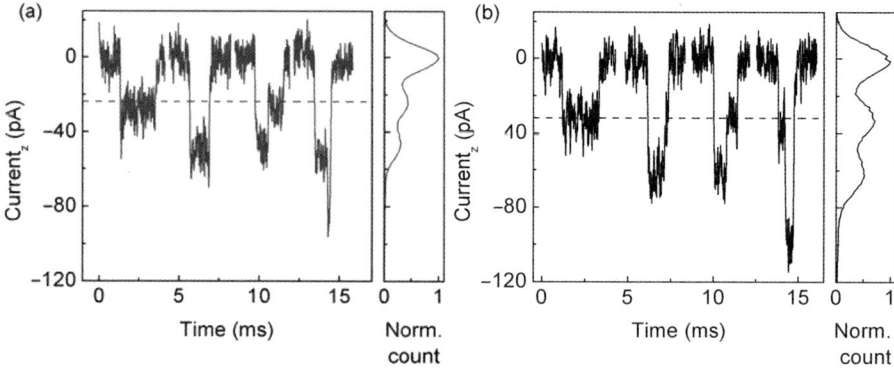

Figure 1 *(a) The left graph shows typical current traces recorded at 0.4 V caused by the translocation of λ-DNA. The baseline current of 12.34±0.24 nA was subtracted resulting in a zeroed current trace, current$_z$. The translocation of λ-DNA caused current decreases with clear quantization visible in the right histogram. This distribution was fitted with a Gauss function revealing a center for the unfolded DNA at 23.73±$0.09 pA marked with a green dashed line. (b) Left graph shows typical current traces recorded at 0.5 V. Again, the baseline current of 15.48±0.03 nA was subtracted. Like in figure (b) fitting the right histogram with a Gauss function revealed a center at 31.93±0.13pA for the unfolded DNA (black dashed line).*

The procedure of determining the I_{un} value was repeated for a potential, U, of 0.3 V. The obtained absolute I_{un} values are plotted in figure 2(a). A clear linear trend is observed which was fitted with a linear function revealing a slope of 67.0 pA/V and an intercept at -2.3 pA at a potential of 0 V. Next, the blocked conductance ($\Delta G_{un} = |I_{un}| / U$) is calculated, which is expected to stay constant. This is what can be observed in figure 2(b) with ΔG_{un} values scattering between 58 and 64 pS. The amplitude of ΔG_{un} does not depend on the applied potential but only on the diameter of the molecule, d, the KCl concentration dependent conductivity, g(c), the sensing length, l_s, the effective charge per unit length, μ_K, and their mobility on the DNA, $q_{l,DNA}$:

$$\Delta G_{un} = \ [\mu_K \ q_{l,DNA} - d^2 \ g(c) \ 4/\pi] \ / \ l_s \qquad (1)$$

A detailed derivation of this equation can be found in recent publications, where it is analyzed as a function of the KCl concentration.[17] Calculating ΔG_{un} using equation (1) for a KCl concentration of 0.5 M and a sensing length of approximately 180 nm results in 40.3 pS.[18] This is in good agreement with the experimental result (mean ΔG_{un} = 61.5±0.2 pS) taking into consideration that the sensing length varies for every nanocapillary.

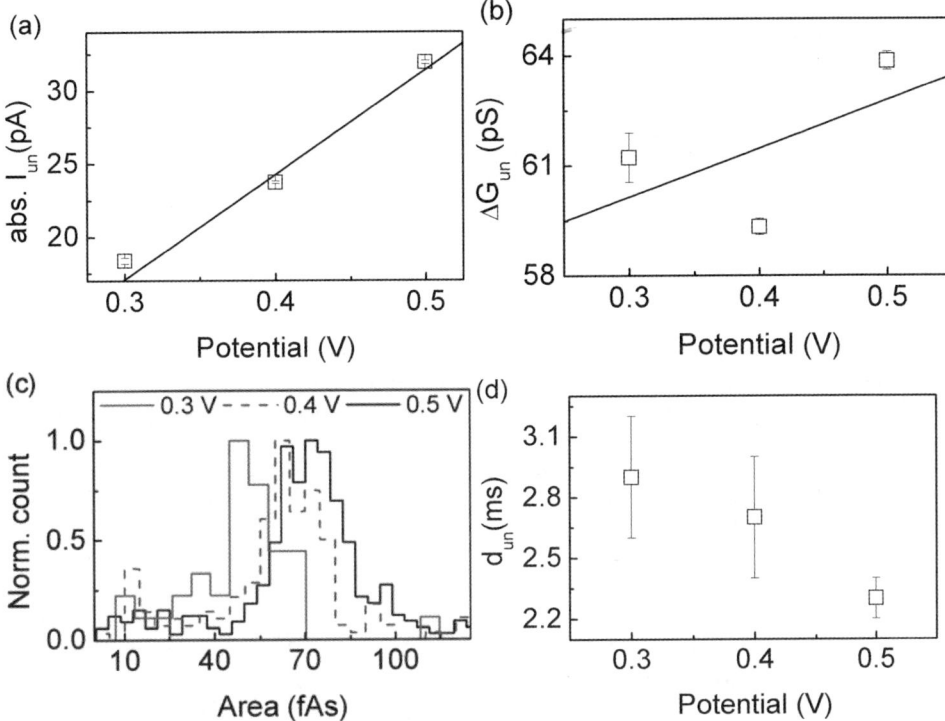

Figure 2 *(a) I_{un} as a function of the applied potential. The dependence was fitted with a linear function revealing an intercept and slope of -2.3 pA and 67.0 pA/V, respectively. (b) The ΔG_{un} value (ΔG_{un} =$|I_{un}|/U$) as a function of the applied potential reveal scattering in the range between 58 and 64 pS. A linear fit showed a subtle decrease for smaller potentials with a slope of 17.7 pS/V. (c) Histogram of the event area normalized to 1 for better comparison. The distributions were fitted with a Gauss function revealing a center at 52±6 fAs, 65±6 fAs and 72±4 fAs for the events at 0.3, 0.4 and 0.5 V, respectively. (d) Duration value for unfolded DNA events, d_{un}, plotted as a function of the applied potential. A clear trend to longer durations can be observed for smaller potentials.*

A linear fit to the distribution in figure 2(b) results in a slope of 3.5 pS over the range of 0.2 V. This represents a change of only 5% relative to measurement at 64 pS. Similar observation were reported by Skinner *et al.* and explained by distortion of the molecular structure or alterations in the access resistance.[14]

Next, we calculated the area, A, of the events at 0.3, 0.4 and 0.5 V. The histograms are shown in figure 5(c). Each histogram was normalized to 1 for better comparison. Again distribution was fitted with a Gauss function to find the center yielding an increasing area value with higher potentials: 52±6 fAs, 65±6 fAs and 72±4 fAs for the experiments at 0.3, 0.4 and 0.5 V, respectively. The folding of the DNA does not affect the amount of excluded charge. In contrast, increasing potentials causes the area to increase as seen in figure 5(c) and the center of the Gauss fit. This shows, that the increase in the I_{un} values dominates in comparison to the decrease in the duration, which becomes smaller for higher potentials.[16]

By dividing the area with the I_{un} one can calculate the duration value, d_{un}, of unfolded DNA (A/ I_{un} = d_{un}). This yielded durations of 2.9±0.3, 2.7±0.3 and 2.3±0.1 ms for the events at 0.3, 0.4 and 0.5 V which are displayed in figure 5(d). It demonstrates that a high potential leads to a short duration whereas a small potential causes longer duration. The decreased electrophoretic force results in longer translocation times and was reported by other groups.[8,16] Although this is obvious it is an additional proof that the observed current peaks, are indeed caused by DNA translocating through the glass nanocapillaries.

3 CONCLUSION

DNA was translocated through a nanocapillary by applying an electrical potential of 0.3 V, 0.4 V and 0.5 V. The translocation duration of unfolded DNA did decrease from 2.83 ms to 2.25 ms when increasing the potential from 0.3 V to 0.5 V. The blocked amount of current increases linearly with the applied potential. Moreover, we demonstrated that the conductance blocked by the DNA stays constant in a range between 59 to 64 pS over a potential ranging from 0.3 V to 0.5 V. These experiments show, that glass nanocapillaries are a cost-effective alternative for DNA translocation through solid state nanopores.

References

1　J. Kasianowicz, E. Brandin, D. Branton, and D. Deamer, *PNAS*, 1996, **93**, 13770.
2　J. Li, D. Stein, C. McMullan, D. Branton, M. Aziz, and J. Golovchenko, *Nature*, 2001, **412**, 166.
3　C. Dekker, *Nature Nanotechnology*, 2007, **2**, 209.
4　L. Ma and S. L. Cockroft, *Chembiochem*, 2010, **11**, 25.
5　G. Maglia, M. R. Restrepo, E. Mikhailova, and H. Bayley, *PNAS*, 2008, **105**, 19720.
6　L. Movileanu, S. Cheley, S. Howorka, O. Braha, and H. Bayley. *J. of General Physiology*, 2001, **117**, 239.
7　M. Faller, M. Niederweis, and G. E. Schulz, *Science*, 2004, **303**, 1189.
8　P. Chen, Y.-R. Kim, J. Gu, E. Brandin, Q. Wang, and D. Branton, *Nano Letters*, 2004, **4**, 2293
9　L. J. Steinbock, O. Otto, DR Skarstam, S. Jahn, C. Chimerel, J. L. Gornall, U. F. Keyser, *Journal of Physics: Condensed Matter*, 2010, **22**, 454113.
10 O Otto, L. J. Steinbock, D. W. Wong, J. L. Gornall, U. F. Keyser, *Review of Scientific Instruments*, 2011, **82**, 086102.
11 C. C. Harrell, Y. Choi, L. P. Horne, L. A. Baker, Z. S. Siwy, and C. R. Martin, *Langmuir*, 2006, **22**, 10837.
12 M. Wanunu, J. Sutin, B. McNally, A. Chow, and A. Meller, *Biophysical J.*, 2008, **95**, 4716.
13 M. Wanunu, W. Morrison, Y. Rabin, A. Y. Grosberg, and A. Meller, *Nature Nanotechnology*, 2010, **5**, 160.

14 G. M. Skinner, M. van den Hout, O. Broekmans, C. Dekker, and N. H. Dekker, *Nano Letters*, 2009, **9**, 2953.

15 L. J. Steinbock, O. Otto, C. Chimerel, J. Gornall, and U. F. Keyser, *Nano Letters*, 2010, **10**, 2493.

16 J. Li, M. Gershow, D. Stein, E. Brandin, and J. A. Golovchenko, *Nature Materials*, 2003, **2**, 611.

17 R. M. M. Smeets, U. F. Keyser, D. Krapf, M.-Y. Wu, N. H. Dekker, and C. Dekker, *Nano Letters*, 2006, **6**, 89.

18 L. J. Steinbock, A. Lucas, O. Otto, and U. F. Keyser, Electrophoresis, 2012, accepted.

TOWARDS SIMULTANEOUS FORCE AND RESISTIVE PULSE SENSING IN PROTEIN NANOPORES USING OPTICAL TWEEZERS

O. J. Pambos[1], K. Göpfrich[1], K.R. Mahendran[2], J. L. Gornall[1], O. Otto[1], L. J. Steinbock[1], C. Chimerel[1], M. Winterhalter[2] and U. F. Keyser[1]

[1] Cavendish Laboratory, University of Cambridge, Cambridge CB3 0HE, UK
[2] School of Engineering and Science, Jacobs University, Bremen, Germany

1 INTRODUCTION

Protein nanopores are highly suitable for single-molecule detection. They offer more reproducible, cost effective and well defined structures than solid-state alternatives with architectures often known from x-ray crystallography studies with sub-nanometer precision[1]. Here we present a self-assembling hybrid nanopore system consisting of a protein nanopore embedded in a lipid membrane, supported across the tip of a nanopipette[2]. Here, we show the insertion of *Staphylococcus aureus* toxin α-hemolysin into the supported membrane and the voltage-driven transport of single-stranded DNA homopolymers. Orientation of the nanopipette perpendicular to the optical trapping axis will allow for high resolution force measurements of macromolecular transport through protein nanopores.

2 METHOD AND RESULTS

2.1 Hybrid Nanopores on Glass Nanopipettes

All component parts of the hybrid nanopore may be formed rapidly and inexpensively on demand using bench-top laboratory instruments. Nanopipettes with tip diameters of approximately 500 nm were formed using a laser pipette puller (Sutter P-2000B) from 0.5mm diameter borosilicate glass capillaries. Giant Unilamellar Vesicles (GUVs) were produced by electroformation in an ITO-coated glass chamber from 10mM DPhPC with 10% cholesterol. 10 µl of the DPhPC solution was deposited onto the ITO glass surfaces with a layer thickness of approximately 500 nm and the chamber was filled with a 1M sorbitol solution. The chamber was maintained at 37.5°C and a 5 Hz sinusoidal potential of $3V_{p-p}$ was applied for a duration of 2 hours. The resulting GUVs had diameters in the range 1-20 µm. Nanopipettes were backfilled with a 300mM KCl buffer and mounted onto a micromanipulator (Figure 1). Ag/AgCl electrodes placed inside the bath solution and nanopipette put the solution in electrical contact with the headstage of a patch-clamp amplifier (Axopatch 200B, Axon Instruments) via a pipette holder. GUVs added to the bath solution were drawn onto the nanopipette tip under a pressure difference generated by a manually-operated syringe. As GUVs came into contact with the nanopipette tip they burst forming bilayers across the tip with resistances in the range 10-150 GΩ. Bilayers formed in this way assemble within seconds, remain stable for up to 10 hours and may be replaced up to 50 times by applying alternating positive and negative pressure with the syringe[2]. Water soluble α-hemolysin (Sigma) was inserted into these supported bilayers by direct addition to the bath solution (Figure 2a). Figure 2a shows the step-wise increase of the ionic current due to insertion of single α-hemolysin nanopores into the bilayer (applied voltage 50 mV). The histogram shows an average steps-size of 12.1pA, as expected from literature[3,4]. The stability of the hybrid nanopore in the presence of a high-power trapping laser was tested by coupling a 500mW Yb-Fiber laser beam (YLM-5-LP, IPG photonics) to the back of a water immersion objective (UPlanSApo/IR, 60×, NA1.2, Olympus)[5] and focusing it onto the nanopipette tip. Figure 3 shows no increase in noise or destabilization of the hybrid pore in the presence of the laser.

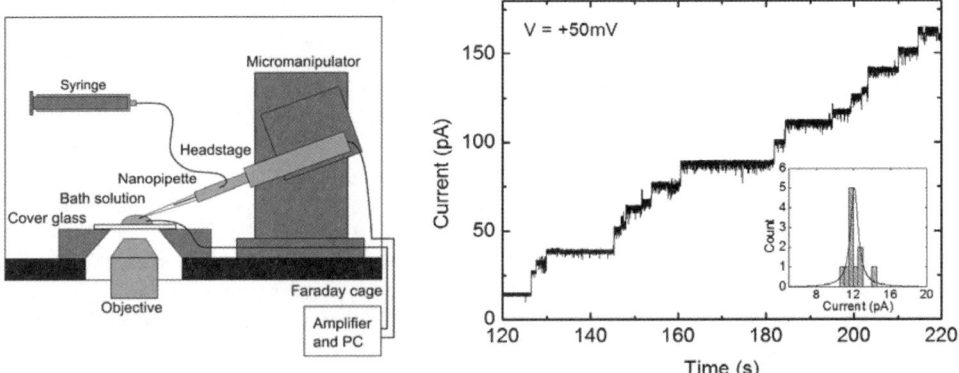

Figure 1 *Left: Experimental set-up. Diagram reproduced from Gornall et al.[2] Right: Multiple insertions of α-hemolysin nanopores observed at with a transmembrane potential of +50 mV*

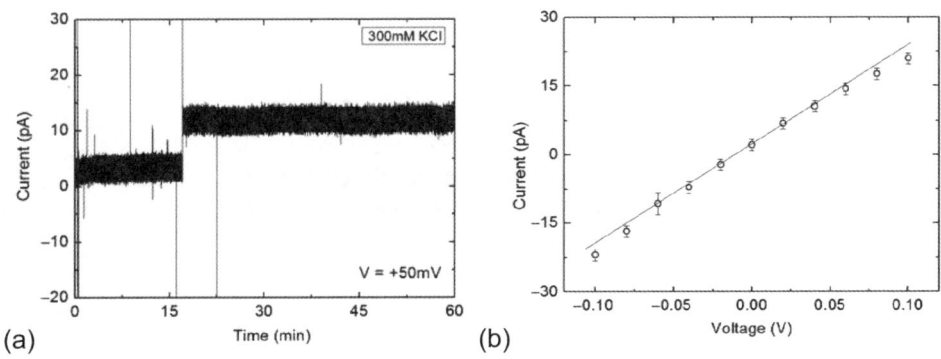

Figure 2 *α-hemolysin in supported membranes. (a) Multiple insertions of α-hemolysin nanopores observed at with a transmembrane potential of +50 mV, (b) Presence of an individual α-hemolysin in a stable membrane for over 40 minutes, (c) I-V characteristics for a single α-hemolysin shows ion current rectification with an ICR ratio of 1.04 at ±100mV, (d) detection of voltage-driven transport of poly(dA)₅₀ ssDNA homopolymers under an applied potential of +50mV. All measurements were carried out in 300mM KCl pH 5.9.*

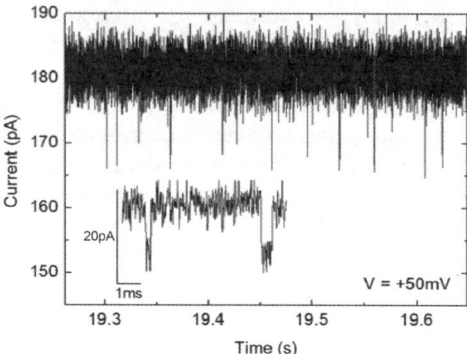

Figure 3 *Detection of voltage-driven transport of poly(dA)$_{50}$ ssDNA homopolymers under an applied potential of +50mV. All measurements were carried out in 300mM KCl pH 5.9.*

2.2 Reconstitution of individual α-hemolysin

Reducing the starting concentration to of 20 µg/ml, a single α-hemolysin nanopore was inserted into a preformed plain supported bilayer (Figure 1). Further insertions were avoided by dilution of the bath solution. The presence of a single nanopore in the membrane was confirmed by conductance measurements (Figure 2a). Ion current rectification was observed in the presence of a single α-hemolysin nanopore with current ratio at ±100mV of 1.04 (Figure 2b), as expected[6].

2.3 DNA translocation measurements

Single-stranded DNA homopolymers of poly(dA)$_{50}$ (Invitrogen) were detected during translocation through supported α-hemolysin nanopores under an applied potential of +50 mV (Figure 4). Observed blockades had a mean duration of ~80 µs, in agreement with published values[7].

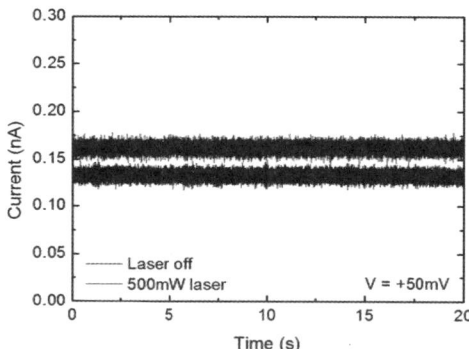

Figure 4 *Stability of hybrid nanopore in the absence (bottom) and presence (top) of a 500mW trapping laser. The increase in baseline current may is attributed to an increase in ion mobility caused by localized heating of the bath solution. Measurement carried out in 300mM KCl pH 5.9.*

2.4 Stability of nanopore with laser irradiation

The goal of our studies is to control single molecules translocating a biological nanopores using optical tweezers. It is necessary for these experiments to use a high-power laser in close proximity to the protein nanopore and the lipid bilayer. We tested the stability of the bilayers with a single inserted nanopore and found that laser powers of up to 500 mW can be focussed on the lipid membrane without disturbing the integrity of the lipid bialyer or the nanopore as shown in Figure 4. The change in current level is explained by the laser heating the solution around the nanopore and thus to an increase in ionic current.

3 CONCLUSIONS AND FURTHER WORK

Measurements of transmembrane conductance confirm the presence of individual α-hemolysin pores in the nanopipette-supported bilayer. Detection of single-stranded DNA homopolymers with blockade durations in agreement with previous studies demonstrates the capability of our nano lipid bilayer approach for single molecule detection. Furthermore, our work has also demonstrated the ability to form a stable hybrid nanopore in a fixed PDMS cell in the presence of a high-power trapping laser. This will allow for simultaneous current and force detection of biomolecules during transport through a protein nanopore. In addition, we hope to carry out high resolution force measurement of protein folding.

References

[1] L. Song, M. R. Hobaugh, C. Shustak, S. Cheley, G. Bayley, J. E. Gouaux, *Science*, 1996, **274**, 1859-1866
[2] J. L. Gornall, K. R. Mahendran, O. J. Pambos, L. J. Steinbock, O. Otto, C. Chimerel, M. Winterhalter, U. F. Keyser, *Nano letters*, 2011, **11**, 3334-3340
[3] D. Wong, T.-J. Joon, J. Schmidt, *Nanotechnology*, 2006, **17**, 3710–3717
[4] D. Wendell, P. Jing, J. Geng, V. Subramaniam, T. J. Lee, C. Montemagno, P. Guo, *Nature Nanotechnology*, 2009, **4**, 765-772
[5] O. Otto, L. J. Steinbock, D. W. Wong, J. L. Gornall, and U. F. Keyser
[6] M. Rincon-Restrepo, E. Mikhailova, H. Bayley, G. Maglia, *Nano letters*, 2011, **11**, 746-750
[7] A. Meller, L. Nivon, E. Brandin, J. Golovchenko, D. Branton, *PNAS*, 2000, **97**, 1079-1084
[8] A. Meller, L. Nivon, D. Branton, *Physical Review Letters*, 2001, **86**, 3435-3438;

STIMULI-TRIGGERED PERMEATION OF IONIC ANALYTES THROUGH NANOPORES FUNCTIONALISED WITH RESPONSIVE MOLECULES

S. Nasir*[1,2] M. Ali[1,2] Q. H. Nguyen,[1,2] and W. Ensinger[1,2]

[1] *Fachbereich Material-u. Geowissenschaften, Fachgebiet Chemische Analytik, Technische Universität Darmstadt, Petersenstr. 23, D-64287 Darmstadt, Germany*
[2] *GSI Helmholtzzentrum für Schwerionenforschung, Planckstr.1 D-64291 Darmstadt, Germany*

1 INTRODUCTION

Since life came into existence mankind is making attempts to compete nature to make life more developed, luxurious, facile and protected. No doubt, it is achieved to a great extent because medical science is now able to transplant heart, kidneys and even other parts of the human body. Scientists are still trying to develop artificial nanofluidic devices to understand the gating mechanism, and permselective transport of ionic species across the cell membrane in living organisms.[1-3] In biological systems, the diffusion of ionic species and small organic molecules occurs through the pore and ion channel across the cell membrane. The gating performance of the cell membrane is controlled by majority of the ion channels that open and close in response to changes in the electrical potential difference and/ or by the binding of a chemical analyte to the interior of the ion channel.[4, 5]

The miniaturization of smart nanofluidic porous systems that respond to environmental stimuli is crucial to mimic and exploit the functionality of natural ion channels in biomedical and health sciences.[4, 6, 7] For this purpose, the surface and inner walls of the nanopores, fabricated in solid-state materials, are decorated with a monolayer and/ or polymer brushes of responsive molecules. Upon the application of external stimuli, permeability of the ionic and molecular species and ionic flux across the responsive porous systems can be remotely controlled and tuned at will for desired function.[6, 8-14]

During the recent years, membrane science and technology has drawn enormous attention because of their outstanding performance in manipulating the ionic and mass transport at the nanoscale level.[15] Polymeric membranes produced by ion-track technology have attracted remarkable interest because of their uniform distribution of pores, tuneable pore geometry, and surface chemical modification.[16] In particular, native chemical groups generated on the pore surface during the track etching process allow the subsequent introduction of responsive molecules/ or polymer brushes to change the transport properties of the membrane.

Here, we will describe the design and functioning of two nanofluidic systems based on arrays of nanopores fabricated in heavy ion tracked polymer membranes, that respond to soft ultraviolet (UV_{365}) light, and environmental temperature (separately). For the case of photosensitive membranes, the inner pore walls are decorated with photolabile 4-oxo-4-(pyren-4-ylmethoxy) butanoic acid (PYBA)[17, 18] protecting groups. Subsequently, the hydrophobic pyrene moieties are removed by UV light irradiation, leading to the exposure of hydrophilic groups, which are responsible for the permselective transport of ionic molecules across the responsive membrane. For the case of temperature responsive nanofluidic system, the inner pore walls are chemically modified with amine-terminated poly(*N*-isopropylacrylamide) [PNIPAAM–NH$_2$] chains *via* "grafting to" approach[19] through carbodiimide coupling chemistry. The effective nanopore diameter is tuned by manipulating the environmental temperature due to the swelling/shrinking behaviour of polymer brushes attached to the inner nanopore walls, leading to a decrease/ increase in the permeation of charged molecular species across the membrane. This process should permit thermal gating and controlled release of ionic drug molecules through the nanopores of the membrane.

2. METHOD AND RESULTS

Commercially available polyethylene terephthalate (PET) membranes of thickness 12 μm were irradiated with swift heavy ion of energy 11.4 MeV/u at the *Helmholtz Center for Heavy Ion Research* (GSI, Darmstadt). Then, the ion tracked PET membranes were sensitized with UV light in the presence of air for 10 minutes on each side. The fabrication of the cylindrical nanopore array (5×10^8 pores cm^{-2}) in the polymer membrane was achieved by the chemical etching of the damaged trails caused by the heavy ions along their trajectories via symmetric track-etching process.[20] The cylindrical geometry of the resulting nanopore has been confirmed by the electrochemical deposition of gold wires inside the nanopores.[20] The fabrication of an array of conical nanopores (5×10^7 pores cm^{-2}) was achieved by well-established asymmetric and selective chemical track-etching techniques developed by Apel *et al.*[21] As a result of heavy ion irradiation and subsequent chemical etching, native carboxylic acid (COOH) groups were exposed on the external surface and inner walls of the nanopores. The surface COOH moieties were exploited for the attachment of responsive molecules onto the pore surface.

2.1. Light-gated nanopores

Track-etched PET membranes containing cylindrical nanopore arrays (~20 ± 3nm in diameter) were used for the fabrication of the photosensitive nanofluidic porous system. The native carboxyl (–COOH) groups were covalently coupled with photosensitive PYBA molecules in a two-step reaction process (Scheme 1).[8]

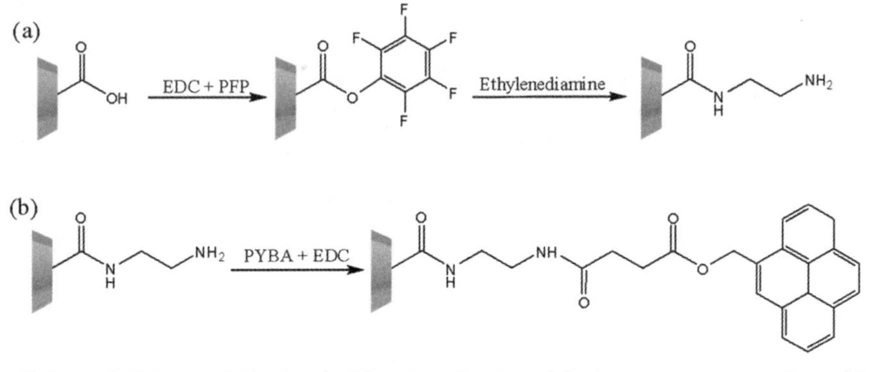

Scheme 1. *Scheme of the chemical functionalisation of the inner nanopore surface with photolabile molecules.*

UV light-induced changes in the pore surface charge and permselective properties were further investigated by mass transport experiments (Figure 1a). It is well known that the ionic transport through the nanopores is mainly governed by the fixed surface charges.[22-25]

The ionic species carrying opposite charge to that of the pore surface will be attracted in and transported selectively through the pore. On the contrary, species with charge of the same sign as that of the surface charge will be repelled and prevented from entering the pore. Neutral pores, however, are non-selective, *i.e.*, both cationic and anionic species are equally transported across the membrane.

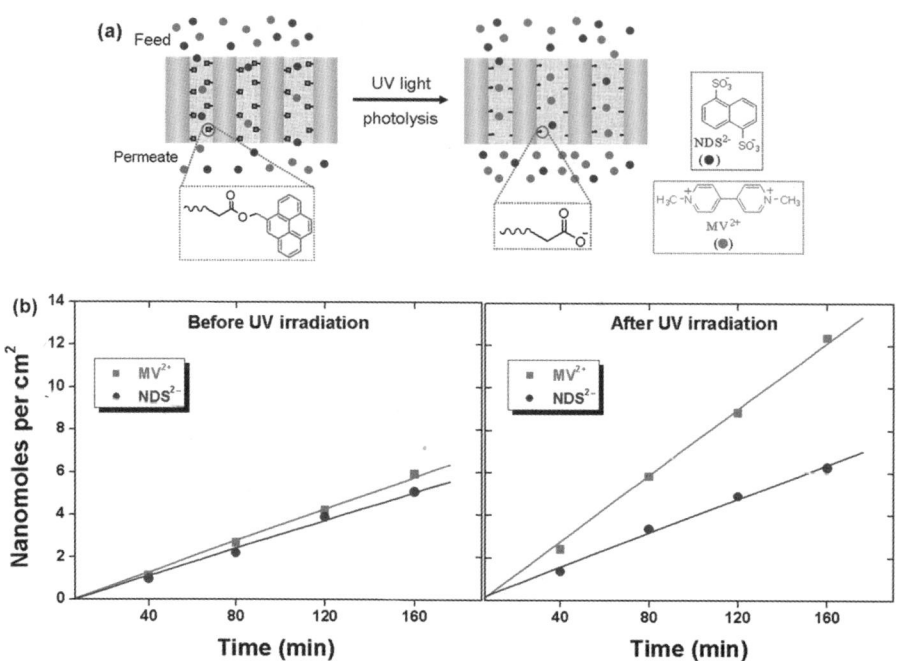

Figure 1. (a) Schematic illustration of the transport of ionic species through the PYBA-modified membrane before and after UV irradiation. (b) Permeation data for MV^{2+} and NDS^{2-} prior to (left) and after (right) UV treatment of the PYBA-modified.

For the transport experiments, PYBA-modified nanoporous membrane was sandwiched between the two compartments of the conductivity cell before and after the UV treatment. The feed compartment was filled with a 10 mM aqueous solution of methylviologen (MV^{2+}) and/ or 1,5-naphthalenedisulfonate (NDS^{2-}) in a phosphate buffer (pH = 6.0), separately. Figure 1b shows the number of moles for the charged analytes MV^{2+} and NDS^{2-} transported per cm^2 of the PYBA-modified membrane *versus* time before and after UV light irradiation. Before the irradiation, the inner walls of the multipore membrane are neutral due to the presence of the uncharged photo-labile pyrene moieties (Figure 1a). These moieties were responsible for the absence of permselective characteristics of the membrane pores. The permeation of both MV^{2+} and NDS^{2-} molecules is similar despite their opposite charges (Figure 1b, left) before the UV treatment. After 160 minutes of analyte diffusion, 5.9 and 5.1 nanomoles of MV^{2+} and NDS^{2-}, respectively, were transported through the membrane. However, these quantities were very different after light irradiation (Figure 1b, right). The permeation of MV^{2+} was remarkably increased with respect to the permeation of NDS^{2-}: the light-induced gating led to an increase of ~110% in the number of moles of MV^{2+} transported (from 5.9 to 12.4 nanomoles) while this increase was only of ~23% in the case of NDS^{2-} (from 5.1 to 6.3 nanomoles).

Upon UV irradiation of the PYBA-modified pores, the targeted photo-labile pyrene moieties were removed, leading to exposed carboxylate (–COO⁻) groups[17, 18] as shown in Figure 1a. These moieties transformed the neutral and hydrophobic inner pore walls into the negatively charged and hydrophilic ones. Therefore, UV irradiation acted as an optical gating, allowing the PYBA-modified membrane to select the cationic (MV^{2+}) species over the anionic (NDS^{2-}) ones.

2.2. Temperature–gated nanopores

The temperature-dependent ionic transport through an array of conical nanopores was investigated by decorating the inner walls with amine-terminated poly(*N*-isopropylacrylamide) [PNIPAAM–NH₂] brushes. Because, PNIPAAM brushes exhibit rapid and highly sensitive conformational transitions triggered by temperature changes in the physiological range.[19] For this purpose, the surface COOH groups were first converted into amine-reactive ester molecules by reacting with an ethanolic solution containing a mixture of *N*-(3-dimethylaminopropyl)-*N'*-ethylcarbodiimide (EDC) and pentafluorophenol (PFP).[26] After washing with ethanol, the PFP-ester molecules were subsequently covalently coupled with the terminal amine groups of the PNIPAAM molecules through amide-bond formation (Scheme 2).

Up to now, most of the previous studies have considered the grafting of polymer brushes onto the surface and inner pore walls through the "grafting from" technique.[14, 27-29] The selection of the PNIPAAM–amine and the "grafting to" technique for the preparation of the thermoresponsive membranes is based on the following facts: 1) the attachment of the PNIPAAM chains on the pore surface is very simple and can be achieved under mild reaction conditions, and 2) the properties of the pre-synthesised polymer molecules are well-known compared to those of different brushes grown by other polymerisation techniques.

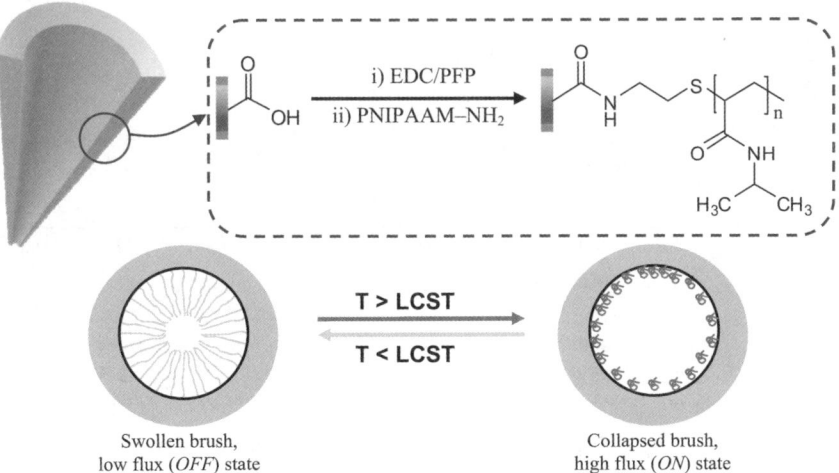

Scheme 2. *Schematic illustration of the amine-terminated PNIPAAM molecules on the inner walls, and thermally-driven conformational transition inside the nanopore.*

Mass transport experiments were performed with PNIPAAM-modified membranes containing an array of conical nanopore. Figure 2 shows the transport of charged analyte molecules as a function of time through the conical nanopores before and after modification of the pore surface with PNIPAAM chains . It is obvious that as-prepared (unmodified) membrane selectively transport MV^{2+} due to the presence of ionised $-COO^-$ groups on the pore surface, while the co-ions (NDS^{2-}) are electrostatically excluded from entering the conical nanopores. A minor increase was observed in the number of the MV^{2+} nanomoles transported across the unmodified membrane when the temperature of the system was increased from 23 to 39°C due to viscosity change of the solution. On the contrary, we did not observe any detectable amount of NDS^{2-} molecules in the permeate-half cell even at high temperature (Figure 2a).

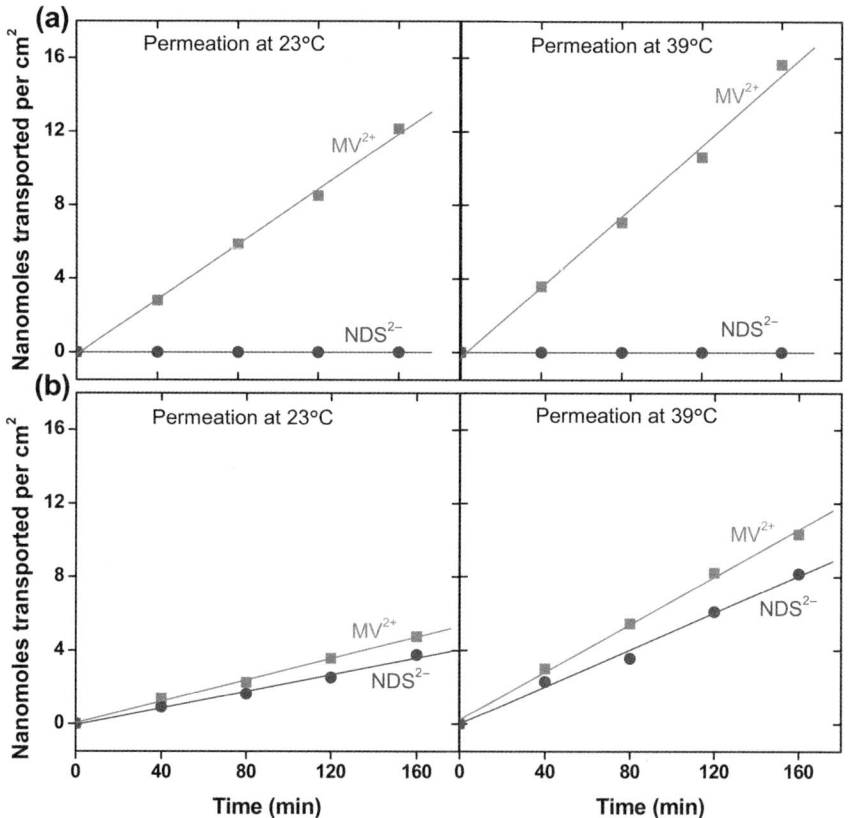

Figure 2. *Temperature-dependent permeation of NDS^{2-} and MV^{2+} molecules prior to (a) and after (b) modification of an array of conical nanopores (5×10^7 pores cm^{-2}) with tip and base openings of $\sim 18 \pm 3$ and 500 ± 5 nm in diameters, respectively.*

Figure 2b shows the permeation data for the case of the modified membrane at low and high temperatures. After functionalisation, the inner walls of the nanoporous membrane (especially the tip region of the conical nanopores) become approximately neutral due to the presence of the uncharged PNIPAAM chains. Therefore, after 160 min of diffusion time, the transport of MV^{2+} (~ 4.7 nanomoles cm^{-2}) and NDS^{2-} (~ 4 nanomoles cm^{-2}) at low temperature is almost the same in spite of their opposite charge

(Figure 2b). At 39°C, the thermal gating yields an increased transport (from 4.7 to 10.3 nanomoles cm^{-2} for MV^{2+} and from ~4 to 8 nanomoles cm^{-2} for NDS^{2-}). The above permeation data constitute a clear evidence that the PNIPAAM chains immobilized onto the nanopore surface undergo conformational changes from a swollen "*OFF*" state (low ionic flux) to a collapsed "*ON*" state (high ionic flux). These changes could be due to the hydration/dehydration of the polymer backbone that occur below/above the lower critical solubility temperature (LCST, ~32°C) of the attached polymer brushes.[19] The conformational changes of the PNIPAAM molecules inside the confined nanopore geometry promote the narrowing or widening of the effective nanopore diameter (pore tip),[14, 27] leading to a decrease or increase in the analyte permeation across the modified-membranes, respectively (Scheme 2).

3. CONCLUSION

We have miniaturized light-gated and temperature-gated nanofluidic systems based on track-etched polymeric membranes *via* decorating the inner channel walls with photolabile (PYBA) molecules and temperature sensitive PNIPAAM chains, respectively. For the case of light-gated nanofluidic system, the hydrophobic terminal pyrene groups in the molecules of attached PYBA can be removed with UV light irradiation, leading to the generation of hydrophilic (COO$^-$) groups and concomitant selective permeation of cationic species through the nanopores across the modified membrane. For the case of PNIPAAM-modified membrane, the effective nanopore diameter available for the permeation of ionic species is externally controlled by the swelling/ shrinking of the polymer brushes attached to the nanopore surface upon manipulating the environmental temperature. We believe that the integration of responsive nanofluidic devices would readily find applications in heat- and light-induced controlled release and permeation of ionic drug molecules at the nanoscale.

References

1. E. Perozo, D. M. Cortes, P. Sompornpisut, A. Kloda and B. Martinac, *Nature*, 2002, **418**, 942-948.
2. S. M. Simon and G. Blobel, *Cell*, 1991, **65**, 371-380.
3. D. A. Butterfield, *Dynamics of biological membranes. by M. D. Houslay, K. K. Stanley, 330 pp, John Wiley & Sons, Inc., New York, NY,* 1983.
4. E. Gouaux and R. MacKinnon, *Science*, 2005, **310**, 1461-1465.
5. B. Hille, *Ionic channels of excitable membranes*, Sinauer Associates Inc., Sunderland, MA, 2001.
6. H. Daiguji, Y. Oka and K. Shirono, *Nano Lett.*, 2005, **5**, 2274-2280.
7. J. Griffiths, *Anal. Chem.*, 2008, **80**, 23-27.
8. M. Ali, S. Nasir, P. Ramirez, I. Ahmed, Q. H. Nguyen, L. Fruk, S. Mafe and W. Ensinger, *Adv. Funct. Mater.*, 2012, **22**, 390-396.
9. M. Ali, P. Ramirez, S. Mafe, R. Neumann and W. Ensinger, *ACS Nano*, 2009, **3**, 603-608.
10. C. Dekker, *Nat. Nanotechnol.*, 2007, **2**, 209-215.
11. X. Hou, W. Guo and L. Jiang, *Chem. Soc. Rev.*, 2011, **40**, 2385-2401.
12. Z. S. Siwy and S. Howorka, *Chem. Soc. Rev.*, 2010, **39**, 1115-1132.
13. I. Vlassiouk, C. D. Park, S. A. Vail, D. Gust and S. Smirnov, *Nano Lett.*, 2006, **6**, 1013-1017.

14.   B. Yameen, M. Ali, R. Neumann, W. Ensinger, W. Knoll and O. Azzaroni, *Small*, 2009, **5**, 1287-1291.
15.   Y. Osada and T. Nakagawa, eds., *Membrane Science and Technology*, Marcel Dekker, Inc, New York, 1992.
16.   R. Spohr, *Radiat. Meas.*, 2005, **40**, 191-202.
17.   W. Cui, X. M. Lu, K. Cui, J. Wu, Y. Wei and Q. H. Lu, *Nanotechnology*, 2011, **22**.
18.   J. Q. Jiang, X. Tong and Y. Zhao, *J. Am. Chem. Soc.*, 2005, **127**, 8290-8291.
19.   B. Zhao and W. J. Brittain, *Prog. Polym. Sci.*, 2000, **25**, 677-710.
20.   Q. H. Nguyen, M. Ali, V. Bayer, R. Neumann and W. Ensinger, *Nanotechnology*, 2010, **21**, 365701.
21.   P. Y. Apel, Y. E. Korchev, Z. Siwy, R. Spohr and M. Yoshida, *Nucl. Instrum. Methods Phys. Res., Sect. B*, 2001, **184**, 337-346.
22.   K. B. Jirage, J. C. Hulteen and C. R. Martin, *Science*, 1997, **278**, 655-658.
23.   C. R. Martin, M. Nishizawa, K. Jirage, M. S. Kang and S. B. Lee, *Adv. Mater.*, 2001, **13**, 1351-1362.
24.   Z. Siwy, E. Heins, C. C. Harrell, P. Kohli and C. R. Martin, *J. Am. Chem. Soc.*, 2004, **126**, 10850-10851.
25.   D. Stein, M. Kruithof and C. Dekker, *Phys. Rev. Lett.*, 2004, **93**, 035901.
26.   M. Ali, B. Schiedt, K. Healy, R. Neumann and W. Ensinger, *Nanotechnology*, 2008, **19**, 085713.
27.   X. Hou, F. Yang, L. Li, Y. L. Song, L. Jiang and D. B. Zhu, *J. Am. Chem. Soc.*, 2010, **132**, 11736-11742.
28.   K. Pan, R. M. Ren, Y. Dan and B. Cao, *J. Appl. Polym. Sci.*, 2011, **122**, 2047-2053.
29.   N. Reber, A. Kuchel, R. Spohr, A. Wolf and M. Yoshida, *J. Membr. Sci.*, 2001, **193**, 49-58.

FABRICATION OF NANOCHANNEL ARRAYS FOR THE SELECTIVE TRANSPORT OF IONIC SPECIES

Q. H. Nguyen,[1,2] M. Ali,[1,2] S. Nasir,[1,2] and W. Ensinger[1,2]

[1] Fachbereich Material-u. Geowissenschaften, Fachgebiet Chemische Analytik, Technische Universität Darmstadt, Petersenstr. 23, D-64287 Darmstadt, Germany
[2] GSI Helmholtzzentrum für Schwerionenforschung, Planckstr. 1, D-64291 Darmstadt, Germany

1 INTRODUCTION

Molecular transport in nanoconfined environment has received considerable interest from the scientific community during recent years. Among these systems, nanoporous polymer membranes have attracted remarkable attention for applications in diffusion, separation and discrimination of organic and biomolecules.[1] The fabrication of nanochannels in solid-state materials (here polymer membranes) was achieved by first irradiating with energetic heavy ions followed by the selective chemical etching of ion tracks. The ion track technology has opened up the possibilities to have a better control over: (1) the channel areal densities, (2) channel size and geometries (various shapes e.g. cylindrical, conical or biconical), and (3) tailoring the surface properties at will to achieve desired interactions with molecules of interest.

Up to now, the most common strategy for tuning the surface properties and opening diameter of track-etched channels in polymer membranes was based on electroless deposition of a thin gold layer on the channel walls followed by the chemisorption of thiolated molecules bearing variable functionalities.[2] Recently, Thayumanavan and co-workers developed a straightforward strategy to control both surface charge properties and effective diameter of channels in polymer membranes. In this methodology, the nanochannel coating was achieved by first adsorption of Sn^{2+} ions on the inner walls followed by the chemical attachment of selected polymer molecules having the desired size and polarity. These modified membranes have been used for the separation and discrimination of small organic and protein molecules based on their charge, size and hydrophobicity.[3]

In this framework, it is highly desirable to develop and design alternative strategies enabling the facile control over channel surface charge by directly exploiting the already existing chemical functionalities on the inner channel walls. In our case, nanochannels fabricated in polyethylene terephthalate (PET) membranes possess carboxylic ($-COOH$) groups were generated on the surface and pore walls during the track-etching process. These groups are negatively charged at physiological pH conditions, due to the presence of ionized $-COO^-$ groups. Furthermore, the surface charge of the nanochannels was reversed by direct covalent linkage of terminated amino ($-NH_2$) groups *via* carbodiimide chemistry. At neutral pH, the protonated amino groups import positive charge to the channel walls.

The nanochannels with fixed surface charges were successfully employed for the selective transport of charged analytes and the construction of molecular gates in which molecular transport can be controlled by manipulating the adenosine triphosphate (ATP) concentration in the external environment.

2 METHOD AND RESULTS

2.1 Fabrication of cylindrical nanochannels

The high fluence irradiation of PET membranes was performed with gold ions having a kinetic energy of 11.4 MeV per nucleon, provided by the linear accelerator UNILAC (GSI, Darmstadt). In order to obtain cylindrical nanochannels, the track membranes were etched by suspending them in a double walled beaker filled with etchant (2 M NaOH) for a preset time. The temperature of the etching solution was maintained at 50 ° C by a circuit of heated water flowing through the double walls of the beaker. After the etching process, the membranes were thoroughly washed with distilled water.

2.2 Functionalization of nanochannels

The –COOH groups on the channel surface were converted into terminated amino groups by the condensation of ethylene diamine (EDA) and/or branched polyethyleneimine (PEI) via *N*-(3-dimethyl-aminopropyl)-*N*-ethylcarbodiimide (EDC)/pentafluorophenol (PFP) coupling chemistry.[4] The transformation of the –COOH groups into amine-reactive esters, a process denoted as activation, was performed in an ethanol solution, containing a mixture of 0.1M EDC and 0.2 M PFP, for 60 min at room temperature. After activation, the samples were immersed in 0.1 M PEI solution, and allowed to react overnight. Finally, the functionalized membranes were washed several times with ethanol followed by deionized water.

2.3 Mass-transport experiments

Unmodified (carboxylated) and modified (aminated) membranes were used for the selective diffusion of charged organic molecules: methyl viologen (MV^{2+}) and 1,5-disulfonate naphthalene (NDS^{2-}). The analyte molecules were dissolved in phosphate buffer saline (10 mM, pH 6.5) solution. For the transport experiments, membranes were mounted between the two halves of aconductivity cell. Each cell volume was 3.4 ml with an effective permeation area of the membrane of 1.15 cm². The feed half-cell contained a known concentration of analyte dissolved in the buffer solution, whereas the permeate half-cell was filled with pure buffer solution. The concentration in the feed half-cell was 10 mM of each MV^{2+} and NDS^{2-} analytes. Both solutions were continuously stirred during the whole transport experiment. After preset time periods, the concentration of analyte in the permeate half-cell was determined by measuring the UV absorbance with a UNICAM UV/vis spectrometer.

2.4 Nanochannel geometry

From the FESEM images of the Au nanowires (Figure 1a) which were electrochemically deposited in the nanochannels as replica, it was obvious that the channels are cylindrical in shape, i.e. they have the same diameter throughout the channel length. This is due to the etching rate along the ion track being considerably higher than the bulk etching rate. The

SEM image of the etched membrane surface (Figure 1b) clearly showed that the channel size distribution is quite narrow.

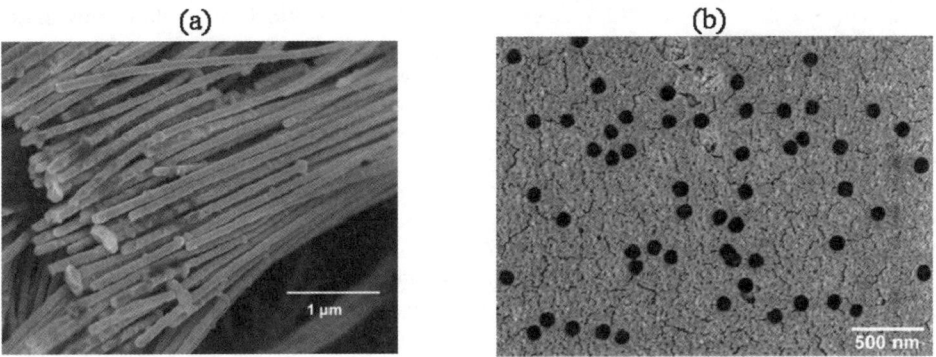

Figure 1 *FESEM images of (a) the Au nanowires deposited in the nanochannels and (b) the surface of the etched membrane.*

2.5 Permeation of charged organic molecules

The permeation flux of charged analytes (MV^{2+} and NDS^2) across the membrane with the density of 3×10^8 channels cm^2 and channel diameter ~20 nm is shown in figure 2. The flux (nanomoles cm^{-2} min^{-1}) reveals the number of moles of the charged molecules transported through the membrane per unit time and is obtained from the slope of these plots. It was observed that in carboxylated (negatively charged) membranes, positively charged analyte (MV^{2+}) has higher flux (1.46) as compared to the negatively charged (NDS^{2-}) analyte (0.19), whereas the aminated (positively charged) membrane selectively transported negatively charged analytes, leading to the increased flux of NDS^{-2} (1.06) and the decreased flux of MV^{2+} (0.58). In general, the unmodified membranes are cation-permselective, while the modified membranes are anion-permselective.

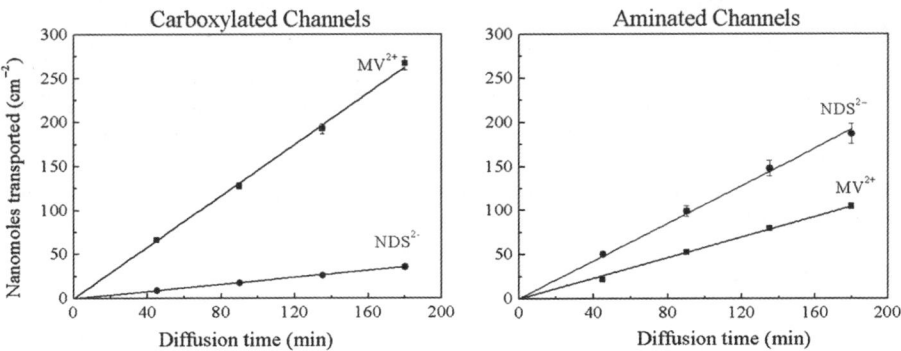

Figure 2 *Diffusion of MV^{2+} and NDS^{2-} through channels with diameter 20 nm prior to (left) and after(right) modification with EDA in membranes with 3×10^8 channels cm^{-2}.*

The driving force for the channel selectivity is the electrostatic interaction between charged walls and analytes. In aqueous solution, the functional -NH_3^+ or –COO^- groups on the channel walls are electrically charged due to protonation or deprotonation, respectively. This charge spurs the formation of an electrical double layer (EDL) inside the channel, which contains a higher concentration of counter-ions than of co-ions. If the diameter of a channel is smaller than the thickness of the electrical double layer, solute ions passing through the pore interact with fixed surface charges. For this reason, charged nanoporous membranes selectively transport ions with opposite charge to the fixed ions on the channel walls, while co-ions are prohibited electrostatically from entering the nanochannel.[5]

2.6 Permeation of biomolecules

Membrane with channel diameter 75 nm was employed for the selective transport of protein molecules such as lysozyme and bovine serum albumin (BSA). The concentration of biomolecules was 10^{-4} M in the feed cell. The flux data of the carboxylated membrane is plotted in figure 3(a), showing that the flux of lysozyme (0.024) was higher than that of BSA (0.01). This is because, in phosphate buffer saline at pH 6.5, BSA (isoelectric point pI 4.7) is negatively charged, and lysozyme (pI 11.0) is positively charged. This also brings electrostatic interactions into play, so that BSA is rejected by the $-COO^-$ layer while lysozyme can pass. Furthermore, due to the higher molecular weight of BSA (66 000) compared to that of lysozyme (14 700), the hydrodynamic radii of BSA and lysozyme are 3.8 nm and 1.8 nm, respectively. As a consequence, the lysozyme molecules can move more freely in the nanochannels which would lead to higher flux of lysozyme. The size effect is obvious when the experiment was performed in ~ 50 nm channels as shown in Figure 3(b). Only lysozyme can be transported across the membrane and no BSA flux was observed.

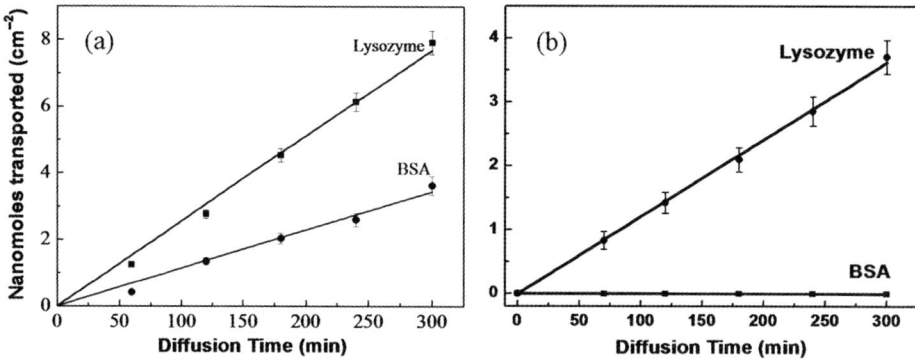

Figure 3 *Diffusion of BSA and lysozyme at pH 6.5 in membranes (3 x 10^8 channels cm^{-2}) with (a) 75 nm and (b) 50 nm channel diameter.*

2.7 ATP-Modulated Molecular Transport

The channel surface was chemically modified with branched polyethyleneimine (PEI) as shown in the scheme of the molecular transport. Figure 4b displays the permeation flux of NDS^{2-} across the membrane before and after the addition of ATP along with analyte in the feed solution. It is evident that in the absence of ATP, the membrane selectively permeates NDS^{2-} molecules across the channels. As mentioned above, the charged channel walls play a significant role in this permeation process. In this case, negatively charged NDS^{2-} ions selectively diffused through the positively charged channels. However, flux of NDS^{2-} was drastically decreased in the presence of ATP in the feed solution (0.57 as compared to 3.03). A possible explanation for this is that the bulky ATP molecules electrostatically attached to the inner walls led to the reduction of surface charges and also inner channel diameter, which in turn suppresses the selectivity and the observed NDS^{2-} ion flux.

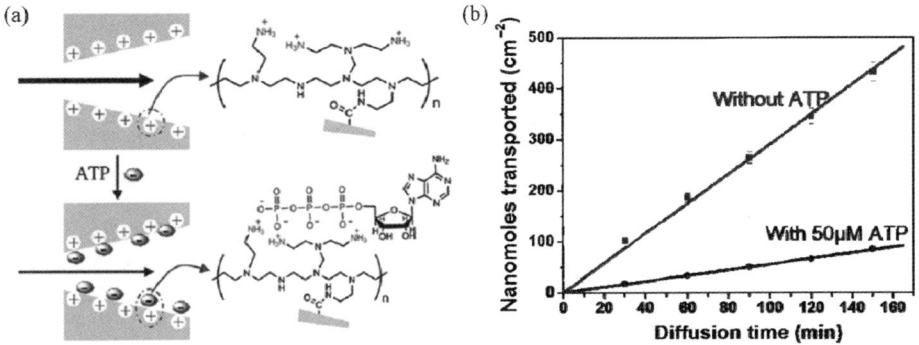

Figure 4 *(a) Scheme illustrating ATP-modulated transport and (b) Diffusion of NDS^{2-} analyte through PEI-modified membrane containing an array of cylindrical nanochannels (10^9 channels cm^{-2}) of ~ 18 nm in diameter, prior to and after the addition of ATP in the analyte solution.*

3 CONCLUSION

In summary, we have described a direct and simple method to control the surface charge properties of ion-tracked PET membranes via carbodiimide coupling chemistry. We have demonstrated that the negatively charged membranes allow the selective transport of positively charged molecules and hinder the passage of negatively charged ones. The selectivity of these filters was reversed by changing the surface polarity from negative to positive via amination. We have also described here the construction of a system which demonstrates ATP-modulated molecular transport through synthetic nanochannels. The molecular flux is chemically controlled via ATP in the surrounding environment. In this context, we believe that this platform would be applicable for the discrimination of enantiomers of drug molecules and multi-responsive drug delivery system at the nanoscale level by introducing suitable functionalities on the channel surface.

ACKNOWLEDGMENT

The authors gratefully acknowledge financial support by the Beilstein-Institut, Frankfurt/Main, Germany, within the research collaboration NanoBiC, and Prof. C. Trautmann for support with the irradiation experiments.

References

1 M. Wirtz, S. Yu and C. R. Martin, *Analyst*, 2002, **127**, 871.
2 C. R. Martin, M. Nishizawa, K. Jirage, M. Kang, and S.B Lee, *Advanced Materials*, 2001, **13**, 1351.

3   E. N. Savariar, K. Krishnamoorthy and S. Thayumanavan, *Nature Nanotechnology*, 2008, **3**, 112.
4   M. Ali, V. Bayer, B. Schiedt, R. Neumann and W. Ensinger, *Nanotechnology*, 2008, **19** 485711.
5   D. Stein, M. Kruithof, and C. Dekker, *Phys. Rev. Lett.*, 2004, **93,** 035901.

GLYCAN ANALYSIS USING A SOLID STATE NANOPORE

M. Takemasa, M. Fujita, and M. Maeda

Bioengineering lab., Advanced Science Institute, RIKEN, 2-1 Hirosawa, Wako, Saitama, 351-0198, Japan

1 INTRODUCTION

Glycans recently attracts attentions as the third biopolymer, next to nucleic acids and polypeptides, since some important roles of glycans were recently clarified due to improvements of analytical methods for glycans. A variety of analysis methods have been developed, but the detailed structure, such as side chain structure in glycoproteins, has not still been accessible especially for relatively high molecular weight glycoproteins or polysaccharides. For example, substitution pattern (attached positions) of side-chains in relation to the length distribution of them varies from molecule to molecule (Fig.1), and these characteristics cannot be determined in the single molecular level by any existing methods such as mass spectrometry, scanning probe microscope and so on.

One of the most difficult points for the analysis comes from variation of these glycan structures. Glycan structure itself is not coded in genome, and it varies in the post-translation modification process in the case of glycoproteins. There are generally several glycosylation sites in single glycoprotein molecule, and the combination of different side chain structures at each site cause each molecule to have different structure in most cases. Currently available analysis methods give us only averaged, and broad information, but it is suggested that these detailed structures are essential for some physicochemical properties and biological activities, which is still not clear due to the lack of effective methods.

The goal of this study is an estimation of (1) attached position and (2) length of each side chain of single glycan molecule, as shown in Fig.1. We are developing analysis system for detailed structure of polysaccharide in the single molecular based on a solid state nanopore.

The applications of nanopore sensing, as nano-scale Coulter counter, are recently expanding from DNA[1] and protein[2] analysis to the other types of molecules, but it has not been reported for glycan structure analysis. Single molecular analysis on glycans has not been performed.

(1) attached position of each side chain

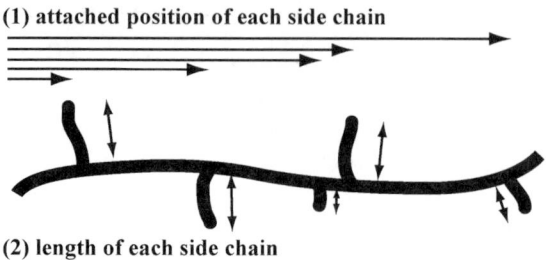

(2) length of each side chain

Figure 1 *Typical structure of glycans, glycoproteins and polysaccharides, can be classified as graft polymer. Characteristic properties of graft polymers, (1) attached position and (2) length of each side chain, are not accessible in the single molecular level by previously reported methods, and the target of this study.*

For the nanopore translocation experiments of glycans, solid state nanopore is required in general instead of biological nanopore to adjust the diameter of glycans, which is much, at least several times, larger than DNA and the diameters of commonly used biological nanopores, transmembrane proteins such as alpha-hemolysin. One of the most advantageous points of solid state nanopore fabricated on thin SiN or SiO$_2$ membrane is adjustability of the pore size[3], in addition to the other characteristics, such as robustness against chemical environment, wide range of pH, solvent type, and so on. Larger pore is required for larger diameter molecules, but smaller pore (closer size to the target molecule) is better for higher S/N ratio. Therefore the optimization is important for the target molecule.

2 METHOD AND RESULTS

Nanopore translocation experiments were performed for polysaccharides carrying charges, such as xanthan and carrageenan, in Tris-HCl aqueous buffer at pH 7.6 in the presence of 1M KCl.

Nanopore was drilled into SiN thin membrane (thickness=30nm, window size=50μm, purchased from TEM windows, West Henrietta, NY, USA) using tightly focused high density electron bean in field emission TEM, JEM-2100F working at 200kV (JOEL, Tokyo, Japan), based on the reported method[3]). The Si/SiN device was set on PDMS microfluidic cell for liquid handling, and transmembrane current was recorded through Ag/AgCl electrodes.

An experimental setup widely used for single channel recording was employed for transmembrane current recordings. Axopatch 200B (Molecular devices, Sunnyvale, CA, USA) was used for current amplifier. Current signal was recorded in Faraday cage with AD board connected to PC, and digitized at 100kHz, 16bit, PEX-361416 (Interface Hiroshima, JAPAN). Internal low-pass filter (4pole Bessel filter, cutoff frequency; 10kHz) was used before acquiring.

Xanthan is extracellular polysaccharide produced by *Xanthomonas campestris*, and is widely used in food and industrial applications. The dried xanthan sample was kindly gift from CP Kelco, and used without further purification.

(a) 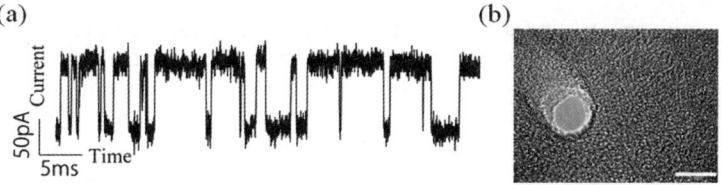 (b)

Figure 2 *Typical nanopore translocation signals of polysaccharides. An aqueous solution of xanthan of 0.1μg/ml in the 1M KCl Tris-HCl buffer was used. Applied voltage=200mV. The size of nanopore = 4nm, and TEM image of the pore was shown in Fig.2b The scale bar represents 5nm.*

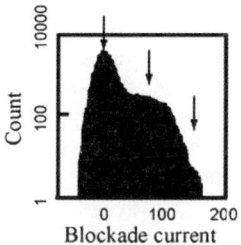

Figure 3. *Current histogram calculated from the transmembrane current. Arrows indicate a peak and shoulders corresponding to the frequently observed current blockade (see text).*

Translocations of polysaccharides were successfully recorded as single molecular events, as shown in Fig. 2. High current level corresponds to open pore, and low one the partial blockade of the pore area by the single molecule polysaccharide translocating the nanopore. Constant amplitude of the pulse indicates that the current blockade events were caused by the single molecular translocation event of the same cross-sectional area of the molecule.

Pulse width reflects the dwell time of each object translocating the nanopore, and averaged dwell time is proportional to the contour length of the polymer, although the pulse width distribution is quite broad even for monodispersed polymer, such as lambda-DNA[4]. The non-constant dwell time of the same length of molecule is caused by the translocation speed across the nanopore is not constant, and the sample used in this study is polydispersed ones.

When the pore diameter is much larger (twice or more) than the diameter of the target molecule for the translocation, multi-level pulses were sometimes observed, which can be interpreted as simultaneous translocation of several molecules at the nanopore position, due to local folded structure of single molecule or several molecules simultaneously translocated into the nanopore, which is essentially the same phenomenon reported for DNA[4]. In other words, internal structure of single molecule polysaccharide is also accessible, and the resolution of the cross-sectional area is quite high, down to single molecular level.

Histogram of the current (Fig.3) showed several peaks (or shoulder), and the current differences between each peak is constant. Most frequently observed current corresponds to the open pore state, and the second shoulder to the blockade current of cross-sectional area of the single molecule. Two times larger blockade amplitude can be interpreted as two molecules simultaneously passed through the nanopore. This is an evidence of the single molecular observation. It was confirmed that the resolution in the cross-sectional area is high enough to distinguish of one glucose level in the cross-sectional area of the single molecule of glycans.

Carrageenan and the other types of polysaccharides carrying charged groups, such as sulfate group, also showed essentially the same types of pulses, indicating that the polysaccharides can also be investigated with solid state nanopore.

3 CONCLUSION

Single molecular translocation events were observed for polysaccharides using a solid state nanopore, and multi-level pulses were observed due to the local folded structure when using larger pore size compared with the diameter of the polysaccharides. These phcnomena are essentially the same as those reported for DNA. This analysis technique, cross-sectional area scanning based on nanopore, is promising not only for glycans but also for branched polymer in more general.

References

[1] Kasianowics JJ, Brandin E, Branton D, Deamer D, *Proc. Natl. Acad. Sci. USA* **93** (24), 13770–3 (1996).
[2] Oukhaled A, Cressiot B, Bacri L, Pastoriza-Gallego M, Betton JM, Bourhis E, Jede R, Gierak J, Auvray L, and Pelta J, ACS Nano, 5 (5), 3628–3638 (2011).
[3] Storm AJ, Chen JH, Ling XS, Zandbergen HW, Dekker C, Nature Mater. 2, 537–540 (2003).
[4] Storm AJ, Chen JH, Zandbergen HW, Dekker C, *J., Phys. Rev. E* **71**, 051903 (2005).

TRANSPORT PROPERTIES OF NANO-POROUS TRACK-ETCHED MEMBRANES IN ELECTROLYTE SOLUTIONS

Andriy Yaroshchuk

ICREA and Department of Chemical Engineering, Universitat Politècnica de Catalunya, av. Diagonal 647, 08028 Barcelona, Spain

1 INTRODUCTION

In **micro-fluidics**, of great importance is the electro-mechanical coupling due to the separation of electrical charges at interfaces (responsible for the electrokinetic phenomena like electro-osmosis). Recently, ever more attention has been paid to **nano-fluidics**. There is a broad range of dimensions where a system already becomes nano-fluidic (e.g., due to the finite thickness of diffuse parts of double electric layers as compared to the channel height), but some of the macroscopic approaches (e.g., continuous description of the solvent) may well remain quantitatively applicable because the system dimensions are still much larger than the molecular scale. Within this range of dimensions, important qualitatively new phenomena occurring in nano-fluidics are related to the concentration-polarization and concentration-gradient-induced phenomena due to mechano-chemical and electro-chemical couplings.

In nano-porous track-etched membranes (TEM), numerous identical "nano-channels" are put in parallel. Therefore, on average, the transport through their ensemble can be considered 1D. For understanding the fundamentals of polarization phenomena at nano/micro interfaces, this one-dimensionality is a clear advantage. Track-etched membranes are obtained through irradiation and subsequent etching of thin polymer films. In this way, systems are obtained having numerous identical straight cylindrical pores. The pore size may range from 10-15 nm to 5 µm depending primarily on the etching conditions.

2 METHOD AND RESULTS

2.1 Experimental

The pressure-induced changes in the concentration of dilute KCl solutions (salt-rejection phenomenon) have been studied with a nano-porous track-etched membrane of *poly (ethylene terephthalate)* (pore diameter ca.21 nm estimated from the hydraulic permeability and known pore density) supplied by Oxyphene in the experimental setup described in [1]. The results are presented in Fig.1.

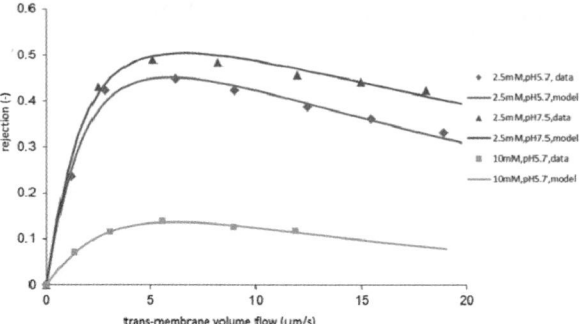

Figure 1 Salt rejection as a function of trans-membrane volume flow. The rejection
decrease at higher flows can be explained by concentration polarization

We have also studied the concomitant stationary trans-membrane electrical phenomenon
(filtration potential), and carried out time-resolved measurements of electrical response to
a rapid (5-10 ms) pressure switch-off [1]. Due to this, the filtration potential could be split
into the streaming-potential and membrane-potential components. An example of such
separation of filtration-potential components is shown in **Figure 2**.

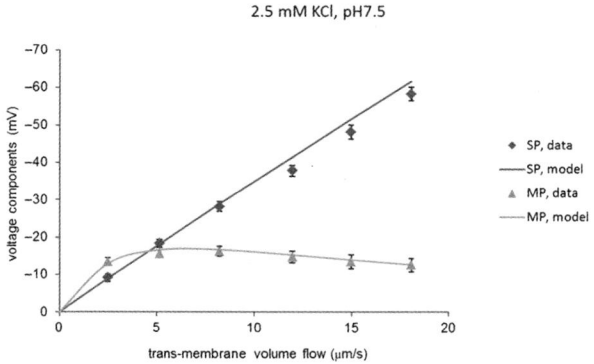

Figure 2 Components of filtration potential obtained via rapid pressure switch-off; SP =
streaming potential, MP = membrane potential.

For the same TEM (clamped between two coarse-porous Millipore filters) and solution, we
have also measured and interpreted transient membrane potential after current switch-off
by using the setup described in [2]. A representative example of corresponding chrono-
potentiogram (and its fitting by the model developed in [2]) is shown in **Figure 3**.

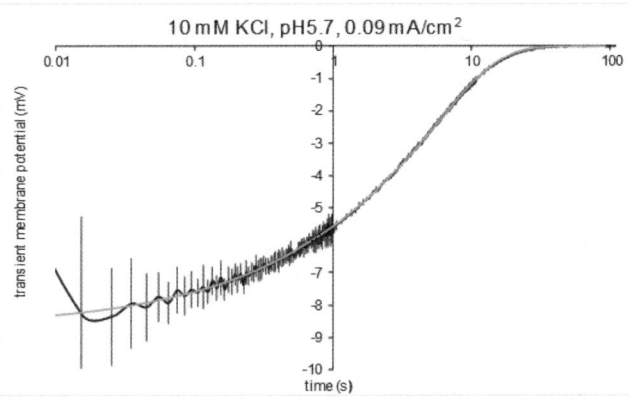

Figure 3 Time transient of membrane potential after current switch-off for a TEM
clammed between two coarse-porous Millipore filters and it's theoretical fit by
the model of [2].

2.2 Results and Discussion

From the theoretical fits of time transients of membrane potential after current switch-off,
we could determine the difference of cation transport numbers between the TEM and
coarse-porous filter. By using the standard space-charge model (see, for example, ref.[3]),
from this difference and the pore size (estimated from the measured membrane hydraulic
permeability and known pore density of $7 \cdot 10^{13}\ m^{-2}$), we could determine the density of
fixed electric charges on the surface of nano-pores. This property could also be estimated
in an alternative way from the salt rejection (Fig.1) and the components of filtration
potential (**Figure 2**). Remarkably, the same values of surface-charge density have been
used to calculate the theoretical fits shown in **Figure 1** and **Figure 2**. Similar fit quality
was observed for the other concentrations and pH values used in the pressure-driven
measurements.

The results of theoretical fitting of surface-charge density (in terms of concentration of
fixed charges) as a function of electrolyte concentration are shown in **Figure 4**.

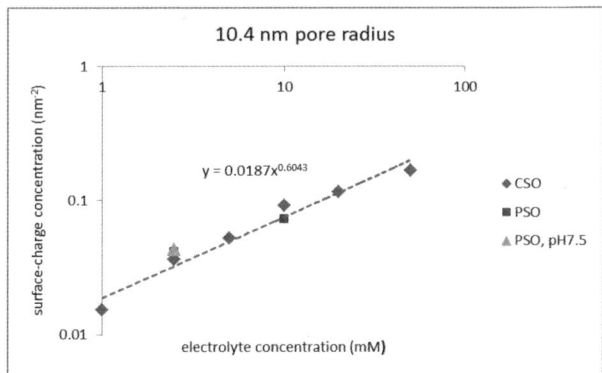

Figure 4 Concentration of fixed electric charges at the nano-pore surface estimated from
the measurements of transient membrane potential after current switch-off and
pressure-driven measurements; CSO = current switch-off, PSO = pressure
switch-off and salt rejection.

The CSO values shown in **Figure 4** are slightly different from those in Fig.11 of ref.[2] due to the use of somewhat different pore size (20.8 nm instead of 24 nm used in [2]).

The data presented in **Figure 4** show a clear trend of increasing surface-charge density with increasing electrolyte concentration. This is in agreement with the probable mechanism of fixed-charge formation due to the dissociation of weakly-acidic surface groups.

It is also remarkable that the CSO and PSO values (where the latter are available) are quite close to each other despite having been estimated from quite different kinds of measurements. Notably, the results of current-switch-off measurements are mostly controlled by simple averages of ion distributions over the nano-pore cross-section whereas the pressure-driven salt rejection and streaming potential are also governed by the weighted averages of those distributions with the (parabolic) profile of convective volume flow through the nano-pores. Therefore, the close agreement between the surface-charge density values estimated from the CSO and PSO measurements is an indication of quantitative applicability of standard space-charge model.

In ref.[4], it has been demonstrated that the pressure-driven transport of electrolyte mixtures with several counterions (ions whose charge is opposite to the fixed charge) can be accompanied by a strong separation of those ions according to their mobilities. In **Figure 5**, we used the value of surface-charge density estimated above from the rejection of KCl and the components of filtration potential in 2.5 mM solution (pH7.5) and the pore diameter of 20.8 nm to predict the membrane performance in the separation of an electrolyte mixture consisting of LiCl (70%) and KCl (30%). Further on, we assumed that the mobilities of isotopes of K^+ are slightly different (0.5%, see ref.[5], for example), and looked at the separation of those two species of very close mobilities. The amplification factor plotted in **Figure 5** is defined in this way

$$amplification\ factor = \frac{selectivity}{relative\ difference\ of\ mobilities}$$

The reciprocal transmission is defined as

$$reciprocal\ transmission = \frac{feed\ concentration}{permeate\ concentration}$$

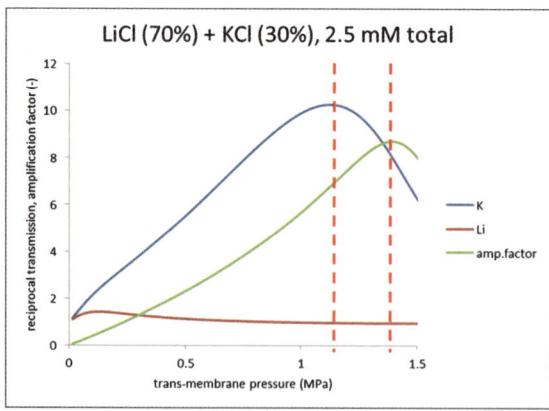

Figure 5 Separation of KCl and LiCl and of isotopes of K^+ in pressure-driven transport through charged nano-pores

First, it is seen that the concentration of KCl is changed up to ca.10 times while that of LiCl remains practically unchanged (very strong separation of K$^+$ and Li$^+$). Second, the amplification factor reaches the value of over 8 at the point of maximum. This means that the selectivity far exceeds the relative difference in the mobilities. This strong separation is the consequence of over-compensation of convective trans-membrane flow of more mobile counterions (K$^+$) by their electric flow in the presence of less mobile ones (Li$^+$) [4].

The values of parameters in the calculations shown in **Figure 5** have been selected to make these estimates correspond to the realistic membrane properties estimated above from several distinct measurements. However, the separation efficiencies featuring in **Figure 5** are far from a possible maximum. This is confirmed by **Figure 6**, which shows the same dependences calculated for a feed solution that is just 2 times more dilute than in **Figure 5**. Moreover, in **Figure 6** we have taken into account the reduction in the fixed-charge density with decreasing concentration according to the empirical correlation represented by the trend line in **Figure 4**. Nevertheless, both the K$^+$ vs. Li$^+$ separation and amplification factor increase considerably as compared to **Figure 5** (note the log scale in **Figure 6**).

An interesting feature of plots in **Figures 5, 6** is the existence of a pressure range where the reciprocal transmission of K$^+$ decreases with trans-membrane pressure but the amplification factor still increases (between the vertical dashed lines). The decreases in both reciprocal transmission and amplification factor are caused by concentration polarization. However, the mismatch in the locations of maxima means that there is a pressure range where an increase in the process productivity (smaller reciprocal transmission means more "product" in the permeate) is accompanied by an increase in the process selectivity. Usually, this is not the case, and one is confronted with a trade-off between productivity and selectivity.

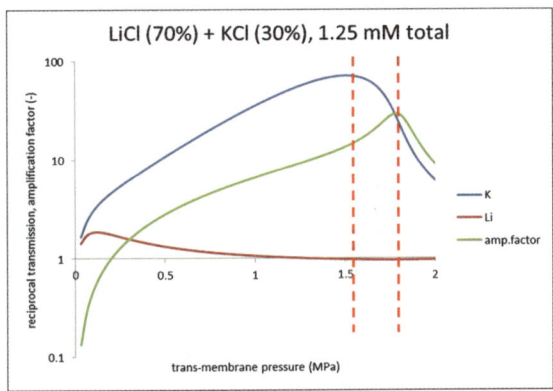

Figure 6 Separation of KCl and LiCl and of isotopes of K in the pressure-driven transport through charged nano-pores; more dilute feed

3 CONCLUSION

Track-etched membranes (TEM) are obtained via irradiation and subsequent etching of thin polymer films. Due to their identical straight cylindrical pores, nano-porous TEMs are suitable for the studies of fundamentals of nano-fluidics. Integrated measurements of salt-rejection, electrokinetic and electrochemical properties of nano-porous TEMs confirm quantitative applicability of standard space-charge model to the description of equilibria and transfer phenomena in nano-pores of ca.20 nm in diameter. In electrolyte mixtures, this same model predicts an interesting phenomenon of strong pressure-driven separation of ions of different mobilities (including very close ones).

References

1. A.Yaroshchuk, Yu.Boiko, A. Makovetskiy *Langmuir*, 2009, **25**, 9605
2. A.Yaroshchuk, O.Zhukova, M.Ulbricht, and V.Ribitsch, *Langmuir*, 2005, **21**, 6872
3. A.Yaroshchuk, *Adv.Colloid & Interface Sci.*, 2011, **168**, 278
4. A.Yaroshchuk, *Adv. Colloid & Interface Sci.*, 1995, **60**, 1
5. S.L. Madorsky, S. Straus, *J.Res.Natl.Bur.Stand.*, 1947, **38**, 185; C.A.Lucy et al, *Can.J.Chem.*, 1999, **77**, 281

SINGLE-MOLECULE DNA TRANSLOCATION THROUGH Si$_3$N$_4$- AND GRAPHENE SOLID-STATE NANOPORES

A. Spiering[1], S. Knust[1], S. Getfert[1], A. Beyer[1], K. Rott[1], L. Redondo[2], K. Tönsing[1], P. Reimann[1], A. Sischka[1] and D. Anselmetti[1]

[1] Faculty of Physics, Bielefeld University, D-33615 Bielefeld, Germany
[2] Institut de Bioenginyeria de Catalunya (IBEC), University of Barcelona, Department of Chemical Physics, Faculty of Chemistry, 08028 Barcelona, Spain

1 INTRODUCTION

The controlled translocation of a single, double-stranded DNA (dsDNA) through a solid-state nanopore (NP)[1,2] with optical tweezers (OT) is described in the presence of an electric field under buffer conditions. Upon threading dsDNA complexed by single proteins through a NP in 20 nm thick Si$_3$N$_4$-membranes, we find distinct asymmetric and retarded force signals that critically depend on the overall charge of the protein, the molecular elasticity of the dsDNA and the counter-ionic shielding of the polyelectrolyte (dsDNA) in the buffer[3]. This force response can be quantitatively simulated and understood within a theoretical model where an isolated charge on an elastic, polyelectrolyte strand experiences a harmonic nanopore potential during translocation. In order to extend these experiments to atomically thin solid-state NP, dsDNA was threaded through single nanolayer graphene NP by a transmembrane voltage. Whereas distinct single-molecule threading signals could here be observed in a Coulter counter setup (Fig. 1a) and compare well with recent papers[4,5], OT controlled dsDNA-translocation through graphene NPs remained challenging, however (Fig. 1b). In this paper, we will give a current status report on optical tweezers controlled translocation through solid-state NPs.

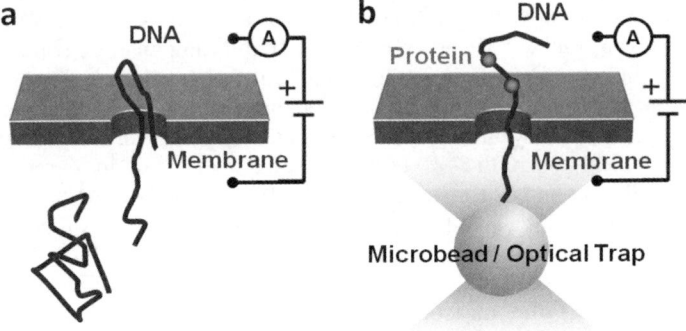

Figure 1 *(a) Coulter counter set-up for single-molecule NP translocation and ionic current detection. (b) Optical tweezers setup for measuring and controlling dsDNA NP translocation.*

2 METHOD AND RESULTS

2.1 Optical Tweezers Setup and Experimental Procedure

We used a single-beam, 3D quantitative OT system with confocal light guiding that incorporates an optical obstruction filter eliminating all axial light components and allowed manipulation and interference-free steering of polystyrene (PS) microbeads under buffer conditions in the vicinity of a reflecting interface (membrane)[6]. Individual Lambda (λ) bacteriophage dsDNA molecules (48502 basepairs (bp), 16.4 µm contour length) were functionalized at one end with several biotins and individually attached to a streptavidin-coated PS bead (3,28 µm diameter). The dsDNA-bead constructs were kept in buffer solution (20 mM / 1M KCl and 2 mM Tris/HCl at pH 8.0) at 22°C and introduced into our NP fluidic cell prior conducting the experiment. In order to probe the force response of an individual protein attached to dsDNA when being threaded through a NP, we introduced either EcoRI (31 kDa) or RecA (38 kDa) proteins.

2.2 Nanopore Fabrication

Solid-state NPs in 20 nm thick Si_3N_4- and single nanolayer graphene membranes were produced by helium-ion microscope milling (Orion, Zeiss, Oberkochen), rendering NP openings with diameters typically 20-60 nm in size.

2.3 Graphene Preparation and Electrical Characterization

Single and multiple graphene nanolayers have been mechanically exfoliated and transferred onto a 20 nm thick Si_3N_4-membrane via the "wedging transfer" technique[5,7] over a 5 m wide hole that was etched into the Si_3N_4-membrane. Since special emphasis has been put on virtually void-free graphene nanolayers for high electrical transmembrane resistance, we used pristine, naturally grown single-crystal graphite nanoflakes (NGS Naturgraphit, Leinburg). In Fig. 2a, a graphene sample with one to three nanolayers is shown in a light microscopy image. A representative current-voltage (I-V) characteristics of a single nanolayer graphene is shown in Fig. 2b, exhibiting an electrical resistance of 29 GΩ in 20 mM KCl buffer solution (Gigaohm seal) as it could only be found in the measured pristine graphene flakes. A 20 nm wide, single nanolayer graphene NP is shown in Fig. 2c. The extraordinary image contrast is due to the He-ion microscope technology that has also been used to drill the NP. In Fig. 2d, an I-V-curve of an 50 nm wide graphene NP is shown, exhibiting an NP resistance of 1,45 MΩ recorded in 1M KCl. In that respect two aspects are worth noting: 1) although control experiments always indicated proper NP fabrication, rather frequently we failed to measure a distinct NP electrical contact resistance even when occasionally surfactants were added to the solution to improve surface wetting, and 2) the measured NP resistance complies well with other data[5] extrapolated for a 50 nm NP.

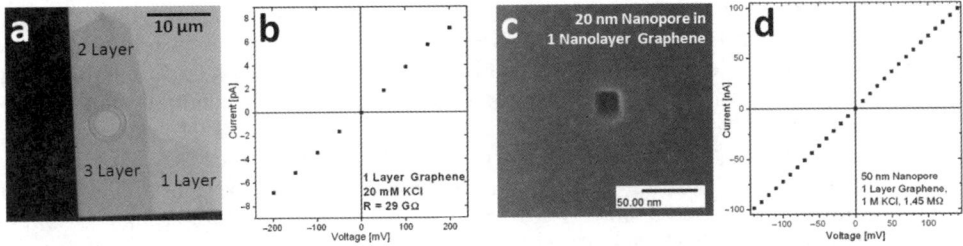

Figure 2 *(a) One to three nanolayer graphene sample transferred to Si₃N₄-membrane. (b) I-V curve of single nanolayer graphene with Gigaohm seal. (c) 20 nm wide NP fabricated in graphene nanolayer by He-ion microscopy milling. (d) 1,45 MΩ contact resistance of 50 nm single nanolayer graphene NP recorded in 1M KCl.*

3 RESULTS AND DISCUSSION

3.1 dsDNA Coulter Counter Experiments

After adding linear λ-DNA to the cis-compartment of our microfluidic cell device[6], and applying an electrical transmembrane voltage (trans: positive, cis: negative) reproducible single-molecule DNA translocation events could be discerned for both Si₃N₄- as well as single nanolayer graphene NPs. In Fig. 3a and b, representative translocation signals are given, reflecting the expected ionic-current blockade during dsDNA translocation for 1 M KCl buffer solution. As already indicated in previous publications, the current variations reflect the folding configuration of the translocating dsDNA-molecules[4,5,8].

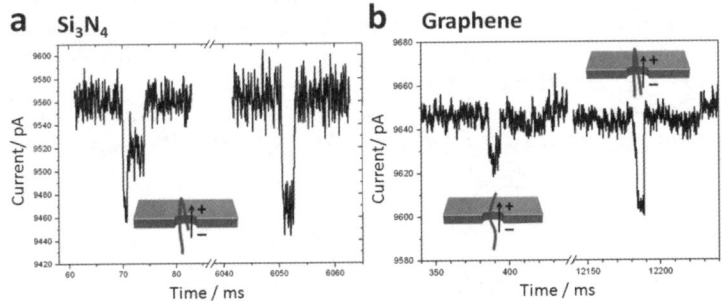

Figure 3 *(a) Coulter counter signals of two λ-DNA single-molecule NP translocation events through a Si₃N₄-NP with a diameter of 40 nm (transmembrane voltage 40 mV). (b) Corresponding signals of λ-DNA single-molecule NP translocation events through a single nanolayer graphene NP with a diameter of 20 nm (transmembrane voltage 140 mV). The current variations are due to the ionic-current blockade and reflect the folding configuration of the translocating DNA.*

3.2 Translocation of Protein-complexed dsDNA with Optical Tweezers (Si₃N₄-NP)

Since we are interested in the quantification of the associated physical mechanisms during molecular translocation, we attached the dsDNA-molecule to a PS-microbead that could be trapped, steered and monitored with our OT setup (Fig. 1b). Upon approaching the NP with a DNA-functionalized microbead and applying a transmembrane voltage (trans: positive) the DNA gets threaded into the NP which can be monitored as a constant force signal. A representative example is depicted in Fig. 4a, where the dsDNA is actively being pulled out of the NP by moving back the microbead with the OT. The jump from the 11 pN background force to zero reflects the threading out of the DNA from the NP. In addition, two characteristic asymmetric force fingerprints (dips) with a retarded force increase extending over more than 200 nm, can be discerned. Each of these asymmetric force signals can be attributed to a single EcoRI-protein that is translocating through the Si_3N_4-NP (Fig 4b). As we could show in an earlier paper[3], this single protein force signal depends on the effective protein charge. Hence and in contrast to the positively charged EcoRI protein, the force response of negatively charged RecA-proteins leads to an inverted force signal (peak), as it is shown in Fig. 4c.

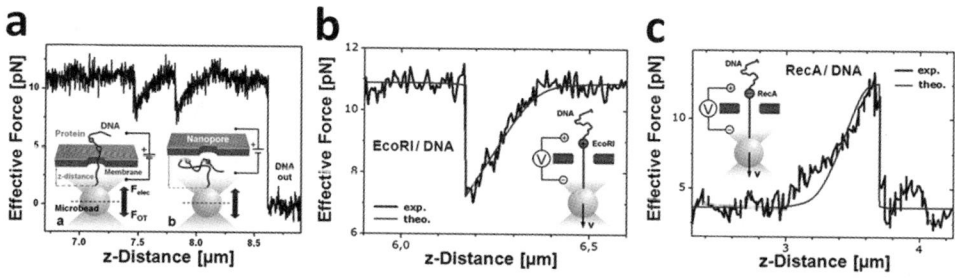

Figure 4 *(a) Force response signal of a protein-complexed λ-DNA-molecule that is being threaded out of a Si_3N_4-NP by OT (45 nm NP diam., 50mV). (b) Close-up of a single-molecule EcoRI/DNA-complex force fingerprint (dip) during NP translocation (45 nm NP diam., 50mV). (c) Force response (peak) of a RecA-complexed λ-DNA-molecule (45 nm NP diam., 20mV). The experimental force response curves (black) in b) and c) were theoretically simulated according a stochastic model[3] (red), yielding the effective protein charge as a result ((b)+59,5e for EcoRI; (c) -700e for RecA-Oligomer) (Figures adapted from Ref. 3).*

The experimental force curves could quantitatively be simulated within a theoretical model where an isolated charge on an elastic, polyelectrolyte strand experiences a harmonic nanopore potential during translocation[3]. As a consequence one finds that the total NP potential for a translocating protein (under external mechanical control) exhibits two minima (potential wells) with a barrier (saddle) in between, corresponding to two metastable "states" with the charged protein on either side of the membrane. As the OT moves, the protein translocates through the NP, however, not in a uniform way but with a dynamics that is governed by thermal fluctuations inducing stochastic transitions between the two states (Fig. 5a). Since these transitions depend on the applied membrane voltage (via the effective NP potential), the measured (averaged) fluctuation time can be related to the activation energy between the two NP states (Kramers rate theory) (see also Fig. 5b).

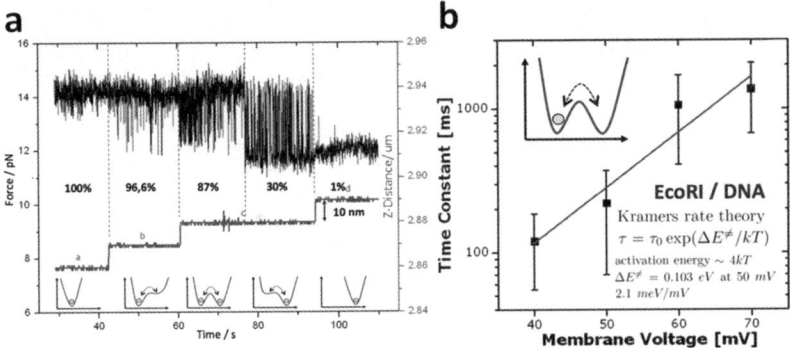

Figure 5 *(a) Thermally activated stochasic transition between the two NP states as can be discerned from the experimental force response curve. The balance between the two-state system can be controlled by the position of the protein along the NP axis coordinate. (b) Estimate of the activation energy between the two NP-states yielding ~4kT at a transmembrane voltage of 50 mV.*

3.3 Towards Translocation of dsDNA with Optical Tweezers (Graphene-NP)

Since the above physical translocation mechanism is governed by non-equilibrium dynamics and depends on the experimental boundary conditions like protein net charge, transmembrane voltage, NP diameter, membrane thickness, ionic buffer strength, electroosmotic currents through the NP (via NP surface charge) and others, it would be insightful to investigate this phenomenon in more detail with other NP geometries and materials. Therefore we chose graphene for this purpose and prepared single nanolayer graphene which we transferred onto Si$_3$N$_4$-membrane supports. The milled graphene NPs with typical diameters of 20-50 nm were tested in Coulter counter experiments (see Figs. 2 and 3), where we monitored individual DNA-passages through the NP.

In contrast, OT-controlled translocation experiments as described in chapter 3.2 for Si$_3$N$_4$-NP turned out to be challenging for graphene NP. The reason might be a systematic difficulty, which stems from the fact that a PS-microbead – optically trapped in a strong IR-laser focus - is obviously heavily thermally activated in the presence of a graphene interface (Fig. 6). This observation suggests that at least the glass transition temperature of PS (around 100 °C) is reached, however, no other phenomena (strong convective flow or vapor microbubbles) can additionally be found. Nevertheless, if this assumption would hold, the integrity of PS-microbead, the dsDNA and its biotin-avidin fixation to the microbead would be questionable[9,10].

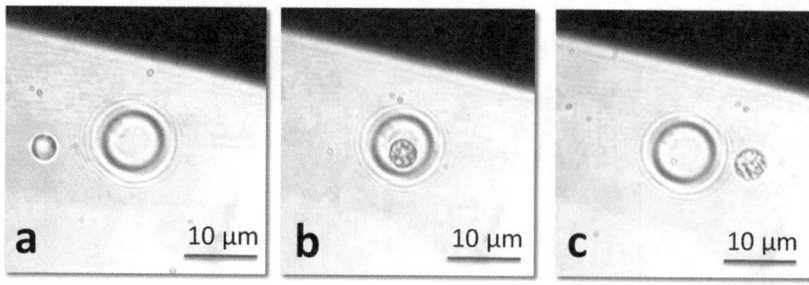

Figure 6 *(a) Video series of light microscopy images, where a PS-microbead (3,28 μm diam.) is positioned with a focussed IR-laser (OT) a few micrometers away from a Si₃N₄-membrane where a single nanolayer graphene is transferred to. The round bore in the middle of the image is a micropore (~10 μm diam.) in the Si₃N₄-membrane window and hosts a free-standing graphene nanolayer (Almost the complete Si₃N₄-membrane is also coated by the graphene). Upon laterally moving the microbead in constant distance across the micropore (with constant laser trapping power of 200 mW), the microbead gets almost instantaneously "heated" only when positioned over the free-standing graphene and turns opaque.*

4 CONCLUSION

The translocation of linear (protein-complexed) λ-DNA through Si₃N₄- and single nanolayer graphene nanopores was investigated with a Coulter counter based setup as well as with optical tweezers control. For the OT controlled translocation experiments through solid-state Si₃N₄-NP, our results provide convincing evidence that force-controlled translocation dynamics of a polyelectrolyte through a NP is accompanied by a thermally induced, stochastic hopping between two adjacent NP states that can adequately be described by Kramers rate theory. Beyond the possibility to detect and reversibly control the position of a single translocating protein attached to a DNA strand by OT, obviously, the overall elasticity of the DNA-polymer significantly contributes to the retarded force response signals when threaded through a NP. It will be interesting to see how force controlled detection concepts will contribute to single molecule NP-based spectroscopy and analysis in the future.

ACKNOWLEDGEMENTS

We would like to thank Dieter Akemeier, Christoph Pelargus, Helene Schellenberg for technical assistance as well as many stimulating discussions with Cees Dekker, Ulrich Keyser, Armin Gölzhäuser and Andreas Hütten. Financial support from the Collaborative Research Centre 613 of the Deutsche Forschungsgemeinschaft (DFG) is gratefully acknowledged.

References

1 U. F. Keyser, B. N. Koeleman, S. van Dorp, D. Krapf, R. M. M. Smeets, S. G. Lemay, N. H. Dekker and C. Dekker, *Nature Physics,* 2006, **2**, 473.
2 A. Sischka, A. Spiering, M. Khaksar, M. Laxa, J. König, K.-J. Dietz, and D. Anselmetti, *J. Phys: Condens. Matter,* 2010, **22**, 454121.
3 A. Spiering, S. Getfert, A. Sischka, P. Reimann, and D. Anselmetti, *NANO Letters*, 2011, **11**, 2978.
4 C. A. Merchant, K. Healy, M. Wanunu, V. Ray, N. Peterman, J. Bartel, M. D. Fischbein, K. Venta, Z. Luo, A. T. C. Johnson, M. Drndic, *NANO Letters*, 2010, **10**, 2915.
5 G. F. Schneider, S. W. Kowalczyk, V. E. Calado, G. Pandraud, H. W. Zandbergen, L. M. K. Vandersypen and C. Dekker, *NANO Letters*, 2010, **10**, 3163.
6 A. Sischka, Ch. Kleimann, W. Hachmann, M. M. Schäfer, I. Seuffert, K. Tönsing, and D. Anselmetti, *Rev. Sci. Instrum.* 2008, **79**, 063702.
7 G. F. Schneider, V. E. Calado, H. Zandbergen, L. M. K. Vandersypen, and C. Dekker, *NANO Letters*, 2010, **10**, 1912.
8 R. M. M. Smeets, U. F. Keyser, D. Krapf, M.-Y. Wu, N. H. Dekker, and C. Dekker, *NANO Letters*, 2006, **6**, 89.
9 M. González, C.E. Argaraña, G. D. Fidelio, Biomol. Eng., 1999, **16**, 67.
10 Y. Li,Y. Fan, J. Ma, Polym. Degrad. Stab. 1999, **65**, 395.

PARALLEL, HIGH-RESOLUTION NANOPORE ANALYSIS ON A CHIP-BASED LIPID MEMBRANE MICORARRAY

Gerhard Baaken,[a,b] , Jan C. Behrends[a,c]

a, *Laboratory for Membrane Physiology and Technology, Department of Physiology, University of Freiburg, Hermann-Herder-Str. 7, 79104 Freiburg, Germany*
b, *Laboratory for Physics and Chemistry of Interfaces, Department of Microsystems Technology, University of Freiburg, Georges-Koehler-Allee 103, 79110 Freiburg, Germany*
c, *Freiburg Centre for Materials Research (FMF), Stefan-Meier-Str. 21, 79104 Freiburg, Germany*

1 INTRODUCTION

In order to be used as single-molecule-sensitive analyte detectors[1-3], biological nanopores need to be functionally reconstituted into free-standing synthetic membranes between two electrolyte compartments that can be contacted by electrodes. Classically, phospholipid bilayers are spread over an opening in a septum between two compartments filled with salt solution and connected by means of non-polarizable electrodes (Ag/AgCl) with the measurement electronics. The set-ups and methods used in current research for such measurements can only be used productively by experienced specialists. They are not amenable to automation and do not support simultaneous measurements at multiple single pores. Future exploitation of the manifold and attractive applications for nanopore analytics in Chemistry and Biology will depend crucially on the development of platform technologies that allow rapid and, if possible, automatic electrical measurements for nanopores with maximally enhanced throughput[4, 5].

A current challenge, therefore, is to create a miniaturized measuring device that allows rapid, reliable and automatable formation of lipid bilayers, protein pore reconstitution and simultaneous measurements at multiple nanopores. In develping such parallel bilayer arays, it is essential that the resolution of the measurements (signal/noise ratio and frequency bandwidth) be uncompromised with respect to state-of-the-art single pore measurements. In fact, miniaturization can and should lead to an improvement over currently used experimental techniques by reducing the electrical capacitance of the entire recording apparatus and set-up. In particular, this involves minimization of the physical dimensions of all electrolyte-wetted surfaces in contact with the amplifier input stage.

2 METHOD AND RESULTS

It is with this intention that in a collaboration between Microsystems Technology and Electrophysiology the MECA (microelectrode cavity array) platform was developed[6, 7]. In contrast to other approaches, where the electrical contact to the nanopore is mediated through electrolyte-filled microfluidic channels that link with the trans- and cis-compartments, the MECA device consists of an array of microstructured cavities in a polymer substrate on glass, with the base of the cavities formed of individual Ag/AgCl-microelectrodes. The cavities have diameters between 6 and 50 μm and a depth of 8 μm (Fig.1 A). They are contacted via coplanar gold conductive paths and can be arranged with a pitch of a few 100 μm, which enables dense integration. Currently we use a prototype with 16 cavities produced using standard microstructuring techniques. Long-term stable Ag/AgCl-microelectrodes with diameters well below 100 μm are a particular challenge that was met with specially tuned silver deposition techniques followed by partial transformation to AgCl.

Measurements on this device using a single-channel high-resolution amplifier showed a noise level that was reduced by a factor of 5 compared to classical systems[7]. This makes it possible, for instance, to produce nanopore-based mass spectra with monomer resolution using an alpha-haemolysin (α-HL) pore[8-10]. Resistive pulses of only 250 μs duration such as are caused by polymers with MW < 1500 g/mol in the pore could be measured precisely and included into the analysis. An example of such an experiment is shown in Fig. 1A. We were also able to perform simultaneous measurements on up to 16 bilipid layers although the multichannel amplifier used for these measurements has much lower temporal resolution (RC=300 μs). Still, with simultaneous measurements on multiple single pores we found single molecule mass spectra with exactly overlapping maxima (Fig. 1B,C). As several groups are currently engaged in developing better multichannel electronics, simultaneous recordings with high time resolution will be possible in the near future.

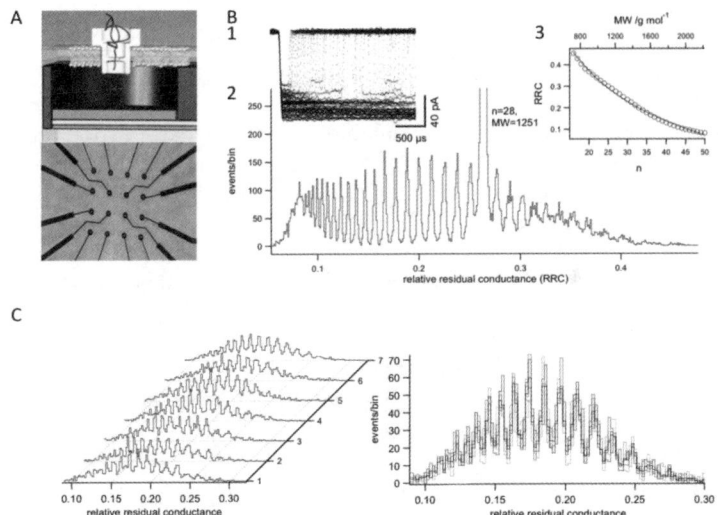

Figure 1 *Polymer Mass-Spectroscopy experiment on the MECA chip. A, Top: Schematic of a single microelectrode cavity with membrane, α-HL pore and polymer. Bottom: Micrograph of central chip surface with 16 cavities; pitch is 200 μm. B: Single-pore recording using a high-resolution single-channel amplifier (Axopatch 2B). Single blocking events (shown superimposed in B1) were automatically detected and the residual conductance measured relative to that of the unblocked pore and displayed as a histogram (B2). The polymer mixture contained PEGs of MW 700-2.200 g/mol to which a monodisperse standard (MW=1.251, n=28) was added, allowing attribution of residual conductance peaks to different polymer sizes[8] (3). C: Histograms obtained in the presence of polydisperse PEG 1.500 using a multichannel amplifier (Tecella Jet). All simultaneously recorded pores show clear maxima (left) the position of which agrees across the different elements of the array.*

3 CONCLUSION

These and other, similarly intentioned developments (for overview see Ref. 4) should enable biological nanopore analysis to become a standard method in Biotechnology and Analytical Chemistry. Planar arrays such as presented here are in principle also well suited for automation of synthetic membrane formation. First experiments in that direction have been promising and[4] there have been advances in understanding and controlling the pore protein reconstition process,[11] justifying the hope that biological nanopore-arrays will prove a first and useful instance of the successful integration of biological nanostructures into technical systems.

References

1.   J. J. Kasianowicz, J. W. F. Robertson, E. R. Chan, J. E. Reiner and V. M. Stanford, *Annu. Rev. Anal. Chem.*, 2008, 1, 737-766.
2.   S. Majd, E. C. Yusko, Y. N. Billeh, M. X. Macrae, J. Yang and M. Mayer, *Curr. Opin. Biotechnol.*, 2010, 21, 439-476.
3.   M. Muthukumar, *Polymer Translocation*, CRC Press, New York, 2011.
4.   S. Demarche, K. Sugihara, T. Zambelli, L. Tiefenauer and J. Vörös, *The Analyst*, 2011, 136, 1077.
5.   S. G. Lemay, *ACS Nano*, 2009, 3, 775-779.
6.   G. Baaken, N. Ankri, A.-K. Schuler, J. Rühe and J. C. Behrends, *ACS Nano*, 2011, 5, 8080-8088.
7.   G. Baaken, M. Sondermann, C. Schlemmer, J. Ruhe and J. C. Behrends, *Lab Chip*, 2008, 8, 938-944.
8.   J. W. F. Robertson, C. G. Rodrigues, V. M. Stanford, K. A. Rubinson, O. V. Krasilnikov and J. J. Kasianowicz, *Proc. Natl. Acad. Sci. USA*, 2007, 104, 8207-8211.
9.   C. G. Rodrigues, D. C. Machado, S. F. Chevtchenko and O. V. Krasilnikov, *Biophys. J.*, 2008, 95, 5186-5192.
10.   J. E. Reiner, J. J. Kasianowicz, B. J. Nablo and J. W. F. Robertson, *Proc. Natl. Acad. Sci. USA*, 2010, 107, 12080-12085.
11.   S. Renner, A. Bessonov and F. C. Simmel, *Appl. Phys. Lett.*, 2011, 98, 083701.

SUBJECT INDEX